D1644623

Textbook

These are the different types of pages and symbols used in this book:

PART ONE

This book is in four Parts. Each Part contains a number of sections relating to different areas of mathematics.

2
Whole numbers 1: Significant figures

These pages develop mathematical skills, concepts and facts in a wide variety of realistic contexts.

53
Extended context

Extended contexts require the use of skills from several different areas of mathematics in a section of work based on a single theme.

47
Detour: Envelopes

Detours provide self-contained activities which often require an exploratory, investigative approach drawing on problem-solving skills.

This symbol shows when you need to use a page from the accompanying workbook.

This is a reminder of key information essential for the work of the page.

Challenges are more-demanding activities designed to stimulate further thought and discussion.

Investigations enhance the work of the page by providing additional opportunities to develop and use problem-solving skills.

Contents

OLYMPIC VIEWING
An estimated 650 million viewers watched the opening ceremony of this year's Olympics, the largest audience for a single event ever recorded.

TRIBUNE SALES ARE UP
Our sales figures for last week soared to a record 6 072 050. The first time in tabloid newspaper history the

Santa's Post
The Post Office announced yesterday that mail over the Christmas period was 60·5 million.

Budget Trimmed
Cine Ltd, the film makers, announced yesterday their intention to cut the cash budget for "Supamaths" by 5·6 million dollars.

Bridge Birthday
A total of 14 560 804 cars have crossed the Tyr Bridge since it was opened on that hot Summer afternoon by the Prince

BENINVER MARATHON
The number of entries for this year's run has risen to 5 762. The organisers are said to be delighted with the response.

SQUIRRELS DOWNFALL
The number of red squirrels has dropped to around 6 500 according to figures released yesterday. The number has fallen steadily for the last few years. Several reasons for the decline were given in the lecture by the naturalist Richard Dunstable.

SOCIETY VOTES
The number of votes cast for the new President of the Arts Society was 50 306 – a staggering figure when compared with the paltry sum needed to gain office last time.

ACCIDENT RATE DOWN
The annual figures for accidents in the home fell dramatically to a new low of 15 694. The statement given out by the Home Office today

Hitting the Jackpot
The winning number in the London Lottery is 160 043 The winner lives near Thurso.

These newspaper cuttings show large numbers. Large numbers are read by grouping the digits in sets of three from the right.

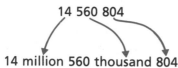

14 560 804

14 million 560 thousand 804

We write this as:
fourteen million, five hundred and sixty thousand, eight hundred and four.

1 Read the cuttings and enter each number **in figures** in the table on **Workbook page 1,** and complete the page.

2 Write, **in words,** the number in each newspaper cutting above.

3 Rewrite this newspaper cutting using **figures** instead of words for the numbers.

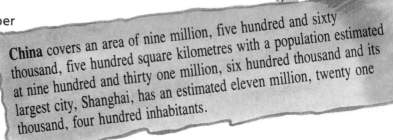

China covers an area of nine million, five hundred and sixty thousand, five hundred square kilometres with a population estimated at nine hundred and thirty one million, six hundred thousand and its largest city, Shanghai, has an estimated eleven million, twenty one thousand, four hundred inhabitants.

EVENING GLOBE
Saturday 26th January

FOOTBALL ATTENDANCES
40 000 at OLD TRAFFORD

Attendances are often approximated in newspaper headlines.

Man Utd v Bolton	43 293
East Fife v Dundee	4947
Queen of Sth v Queen's Park	537
Gartferry v Rigside	

The *Evening Globe* reporter Rob has used the highest place value to approximate the crowd size at Old Trafford.

tens of thousands

At Manchester United **43 293** is **40 000** to the nearest ten thousand.
At East Fife **4947** is **5000** to the nearest thousand.

thousands

These attendances have been **rounded to one significant figure.**

1 Do Workbook page 2, question 1.

2 Round each attendance to one significant figure.

(a)	Forfar v Celtic	8359	**(b)**	Millwall v Sheffield Wed	13 663
(c)	Shotts v Eagle Inn	18	**(d)**	Stenhousemuir v Albion R	497
(e)	Port Vale v Man City	19 132	**(f)**	York v Hartlepool	3075
(g)	Partick v Falkirk	9552	**(h)**	Liverpool v Brighton	32 670
(i)	Dumbarton v Alloa	551	**(j)**	Norwich v Swindon	14 408

The Morning World reporter Jane used the second highest place value to approximate the crowd size at Old Trafford to **two significant figures.**

thousands

At Manchester United **43 293** is **43 000** to the nearest thousand.
The attendance has been **rounded to two significant figures.**

MORNING WORLD
Sunday 27th January

43 000 at BIG MATCH

3 Do Workbook page 2, question 2.

4 Make a list of these matches and round each attendance to
  (a) one significant figure
  (b) two significant figures.

Saturday 16th March

Aston Villa	v	Tottenham	32 638
Sheff Utd	v	Chelsea	20 581
Motherwell	v	Morton	9005
Nottm For	v	Man Utd	23 859
Southmptn	v	Everton	15 410
Arbroath	v	E. Stirling	272

Doug, United's manager, has drawn this plan to show crowd capacity in United's Stadium.

	BUSBY STAND 9480	
STEIN TERRACE 11600		MATTHEWS TERRACE 16400
	ENCLOSURE 5 650	
	RAMSAY STAND 7800	

What is the total capacity of the two stands?
You can find an approximate total by first rounding each number to one significant figure.

9480 is about 9000
7800 is about 8000

The **approximate** total is 9000 + 8000 = **17 000**
The **exact** total is 9480 + 7800 = **17 280**

For each question
- find an **approximate** answer by rounding each number to one significant figure
- find the **exact** answer
- check that both answers are about the same.

1 What is the total capacity of
 (a) the terraces   **(b)** the stands and the enclosure?

2 What is the **difference** in capacity between
 (a) the terraces   **(b)** the stands?

3 Doug is calculating takings from a recent cup game. His records show attendances and ticket prices. Calculate the takings from each section of the stadium.

	Busby Stand	Ramsay Stand	Stein Terrace	Matthews Terrace	Enclosure
Attendance	8982	6012	10 982	15 690	5007
Price	£18	£12·50	£11	£7·70	£5·20

4 For each team calculate the average attendance to
 (a) 31st December   **(b)** 31st May.

	Total attendance to 31st December	Total attendance to 31st May
Albion	42 218 (19 matches)	81 614 (43 matches)
Thistle	28 896 (14 matches)	86 048 (32 matches)
Rovers	16 742 (22 matches)	19 874 (38 matches)
United	422 583 (21 matches)	881 498 (46 matches)

Ask your teacher what to do next.

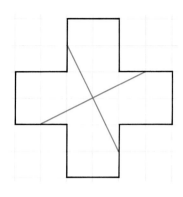

1 • Copy the cross on 1 cm squared paper and cut it out.
 • Cut along the **red** lines.
 • Fit the 4 pieces together to form a square.
 • Draw the square to show how the pieces fit.

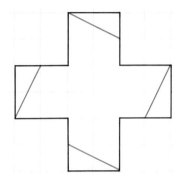

2 • Copy the cross on 1 cm squared paper and cut it out.
 • Cut along the **red** lines.
 • Fit the 5 pieces together to form a square.
 • Draw the square to show how the pieces fit.

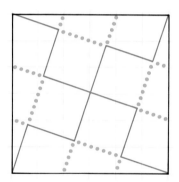

3 • Copy this diagram on 1 cm squared paper and cut it out.
 • Cut along the **red** lines to make 4 congruent pieces.
 • Fit the pieces together to form **two** crosses.
 • Draw a cross to show how the pieces fit.

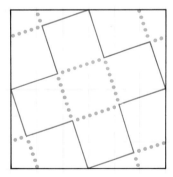

4 • Copy this diagram on 1 cm squared paper and cut it out.
 • Cut along the **red** lines to make one cross and four pentagons.
 • Fit these pentagons together to form another cross.
 • Show how the pieces fit to make the second cross.

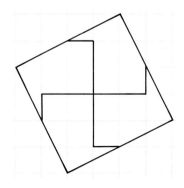

5 • In the middle of a sheet of 1 cm squared paper draw a tile as shown.
 • Extend the pattern in all directions to make a tessellation of identical tiles.
 • Colour any **crosses** you find in the tessellation.

Ask your teacher what to do next.

Sheila designs club logos for Grafix Design. She has to enlarge
some logos to fit different products.

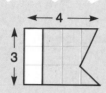

This logo has been
enlarged by multiplying
each length by 2.
The scale factor is 2.

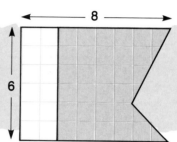

1 Do Workbook page 3.

You need $\frac{1}{2}$ **cm squared paper.**

2 Enlarge or reduce each logo by the given scale factor.

(a) scale factor 2   **(b)** scale factor $\frac{1}{2}$ **(c)** scale factor 4   **(d)** scale factor $\frac{1}{3}$.

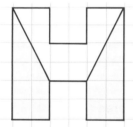 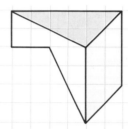 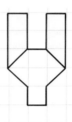 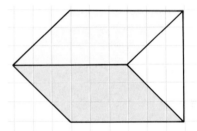

3 (a) Design a logo using your own initials
  (b) Enlarge it by a scale factor of 3.

4 Find the scale factor for each of these.

(a)   **(b)**   **(c)**

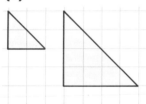 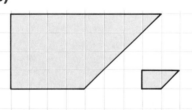 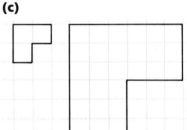

A shape and its enlargement are **similar** to each other.

5 Find pairs of shapes which are similar to each other.

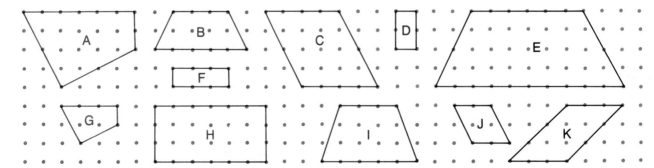

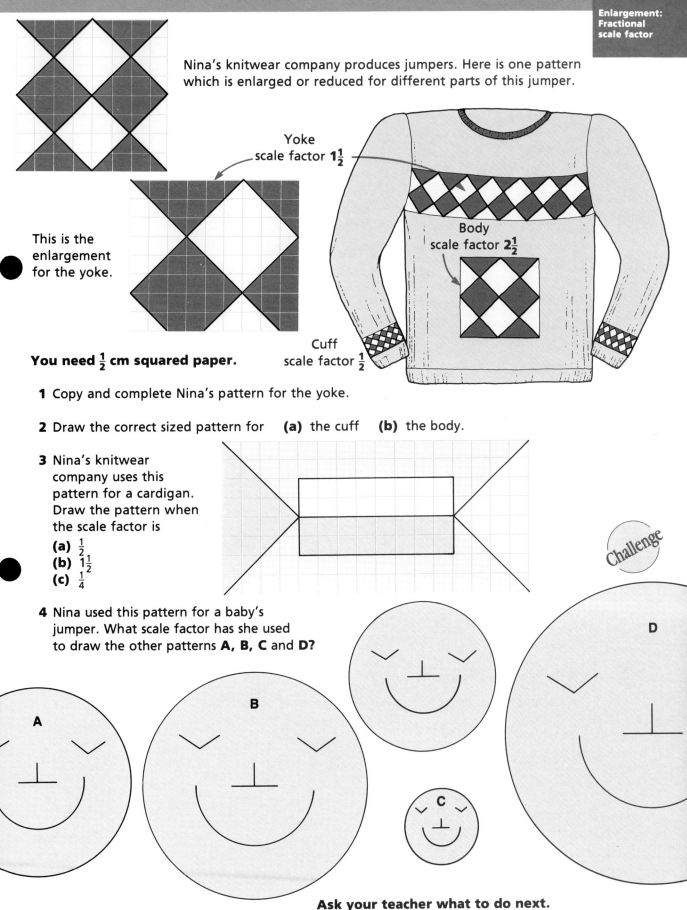

Nina's knitwear company produces jumpers. Here is one pattern which is enlarged or reduced for different parts of this jumper.

Yoke
scale factor $1\frac{1}{2}$

This is the enlargement for the yoke.

Body
scale factor $2\frac{1}{2}$

You need $\frac{1}{2}$ cm squared paper.

Cuff
scale factor $\frac{1}{2}$

1 Copy and complete Nina's pattern for the yoke.

2 Draw the correct sized pattern for **(a)** the cuff **(b)** the body.

3 Nina's knitwear company uses this pattern for a cardigan. Draw the pattern when the scale factor is
  (a) $\frac{1}{2}$
  (b) $1\frac{1}{2}$
  (c) $\frac{1}{4}$

Challenge

4 Nina used this pattern for a baby's jumper. What scale factor has she used to draw the other patterns **A, B, C** and **D**?

A

B

C

D

Ask your teacher what to do next.

Plain sailing

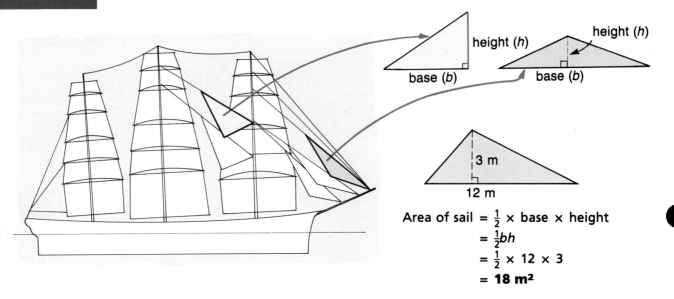

height (*h*)

base (*b*)

height (*h*)

base (*b*)

3 m

12 m

Area of sail = $\frac{1}{2}$ × base × height

= $\frac{1}{2}bh$

= $\frac{1}{2}$ × 12 × 3

= **18 m²**

John, the sailmaker, has to work out the area of each of his sails.

1 Find the area of each of these sails.

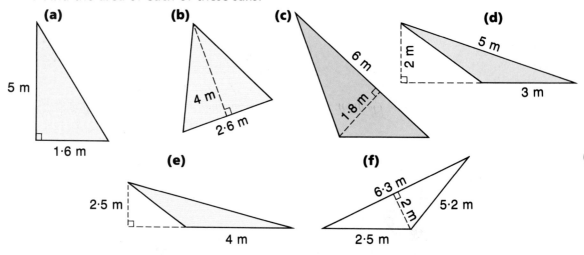

(a)

5 m

1·6 m

(b)

4 m

2·6 m

(c)

6 m

1·8 m

(d)

2 m

5 m

3 m

(e)

2·5 m

4 m

(f)

6·3 m

2 m

5·2 m

2·5 m

2 Do Workbook page 4.

3 Each of these sails is a rhombus. Find the area of each sail.

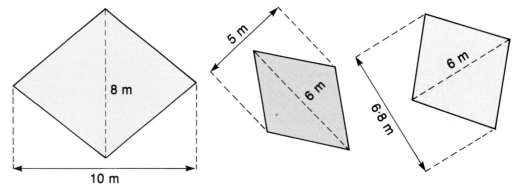

8 m

10 m

5 m

6 m

6 m

6·8 m

4 Do Workbook page 5.

Tom designs tie-pins for the company All Tied Up.
He used these shapes in his new designs.

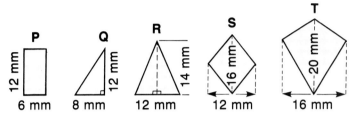

1 Name each shape and find its area.

2 These are some of the designs he made from shapes
which are congruent to the ones above.
For each tie-pin design **(a)** name the shapes used
(b) find the total area.

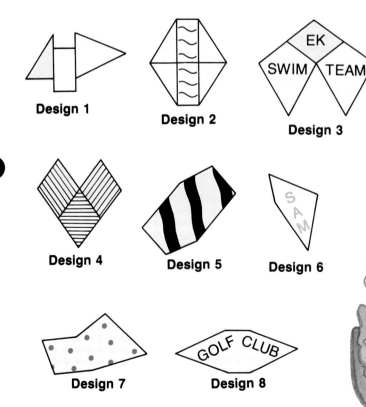

Design 1

Design 2

Design 3

Design 4

Design 5

Design 6

Design 7

Design 8

3 On 1 cm squared paper use geometric shapes to design
some tie-pins. For each of your designs
• name the shapes used   • calculate the total area.

Ask your teacher what to do next.

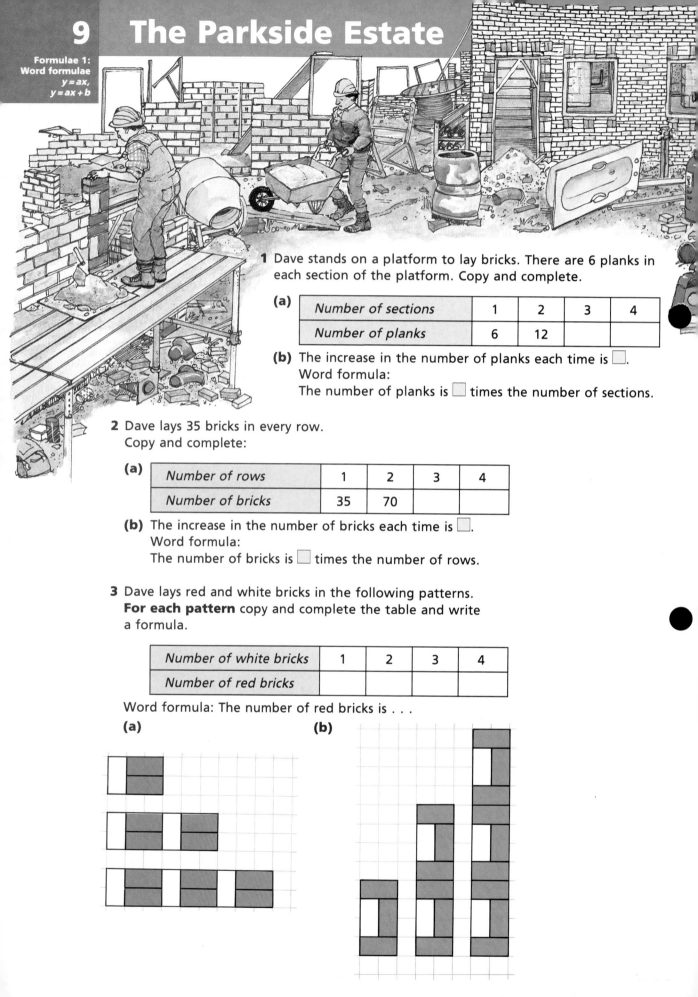

1 Dave stands on a platform to lay bricks. There are 6 planks in each section of the platform. Copy and complete.

(a)

Number of sections	1	2	3	4
Number of planks	6	12		

(b) The increase in the number of planks each time is ☐.
Word formula:
The number of planks is ☐ times the number of sections.

2 Dave lays 35 bricks in every row.
Copy and complete:

(a)

Number of rows	1	2	3	4
Number of bricks	35	70		

(b) The increase in the number of bricks each time is ☐.
Word formula:
The number of bricks is ☐ times the number of rows.

3 Dave lays red and white bricks in the following patterns.
For each pattern copy and complete the table and write a formula.

Number of white bricks	1	2	3	4
Number of red bricks				

Word formula: The number of red bricks is . . .

(a)   **(b)**

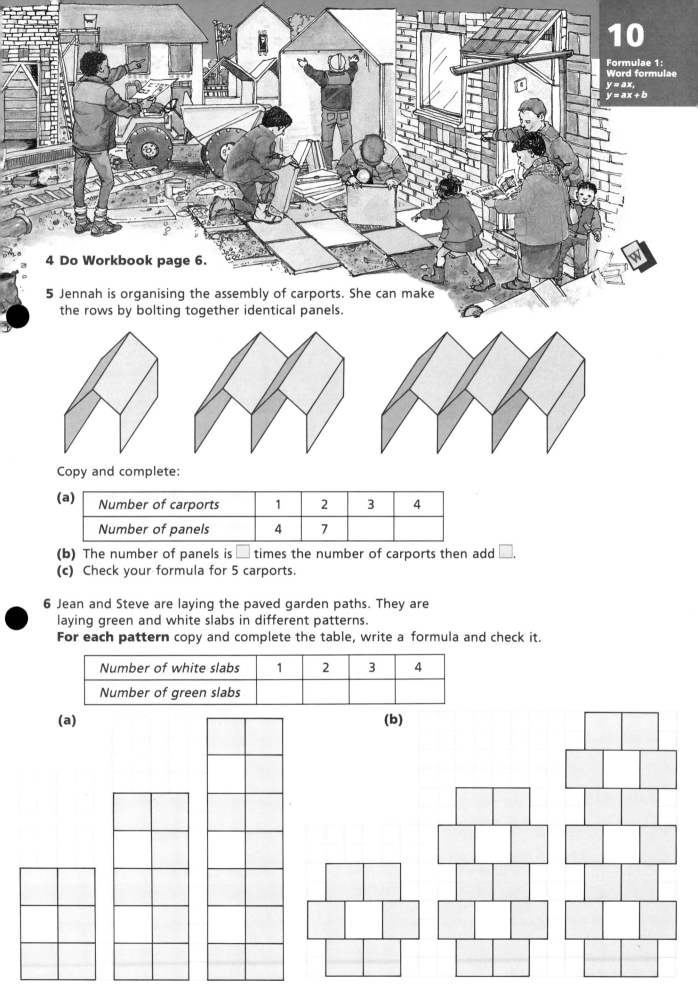

4 Do Workbook page 6.

5 Jennah is organising the assembly of carports. She can make the rows by bolting together identical panels.

Copy and complete:

(a)

Number of carports	1	2	3	4
Number of panels	4	7		

(b) The number of panels is ☐ times the number of carports then add ☐.

(c) Check your formula for 5 carports.

6 Jean and Steve are laying the paved garden paths. They are laying green and white slabs in different patterns.
For each pattern copy and complete the table, write a formula and check it.

Number of white slabs	1	2	3	4
Number of green slabs				

(a)

(b)

Ferrers' catalogue page shows different fence designs made by welding together iron bars.

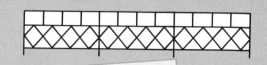

INFINITE VARIETY

1 (a) Copy and complete for the square design.

Square design

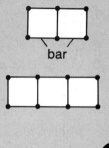

Number of square sections (s)	1	2	3	4	5
Number of iron bars (b)					16

bar

The increase in the number of bars each time is 3.
Number of bars = 3 × number of squares + 1
$b = 3 \times s + 1$
The formula in letters is **$b = 3s + 1$**

Check for 6 squares:
$s = 6$ $b = 3s + 1$
         $b = 3 \times 6 + 1$
         $b = 19$

There are 19 bars in 6 squares.

weld

(b) Use the formula **$b = 3s + 1$** to find the number of bars in   • 10 squares   • 15 squares.

2 (a) Copy and complete.

GUARANTEED NO RUST

Number of square sections (s)	1	2	3	4	5
Number of welds (w)		6			

(b) Write a formula for the number of welds
 • in words   • in letters.
(c) Check the formula for 6 squares.
(d) Use the formula to find the number of welds in
 • 12 squares   • 20 squares.

EASY ASSEMBLY

Diamond design

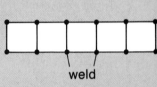

3 (a) How many welds are needed for a fence with
 • 4 diamonds   • 5 diamonds?
(b) Make a table to show the number of welds needed for different numbers of diamond sections.
(c) Write a formula for the number of welds
 • in words   • in letters.
(d) Check the formula for 6 diamonds.
(e) Use the formula to find the number of welds in
 • 13 diamonds   • 20 diamonds.

LONG-LASTING!

4 For each of these fence designs write a formula to find the number of **bars** when you know the number of sections. Check each formula.

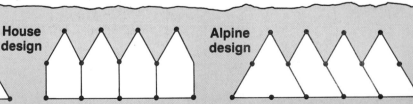

Triangle design   House design   Alpine design

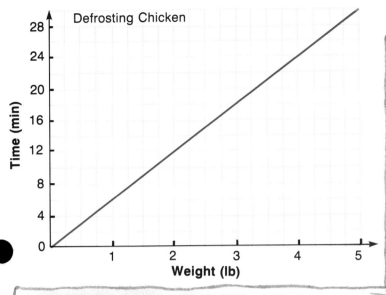

Defrosting Chicken

Sarah is a chef in the Glacier Valley Hotel. She is defrosting frozen food before cooking it. The graph shows microwave defrosting times for chicken.

1 (a) Copy and complete.

Weight in lb (W)	1	2	3	4
Time in min (T)	6			

(b) Copy and complete: Time = ☐ × Weight   or $T = ☐W$
(c) Check that your formula works for a chicken weighing 5 lb.
(d) How long will it take to defrost a 9 lb chicken?

2 For each of these graphs
- make a table
- write a formula
- check your formula.

(a)

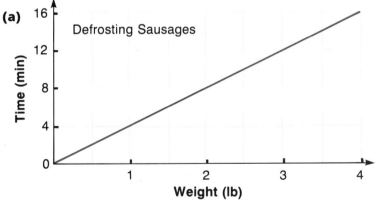

Defrosting Sausages

(b)

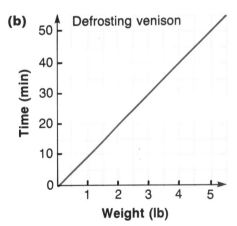

Defrosting venison

(c)

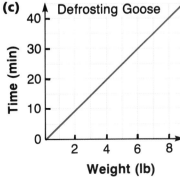

Defrosting Goose

3 How long must Sarah allow for defrosting
(a) 6 lb of sausages
(b) 8 lb of venison
(c) 11 lb of goose
(d) $9\frac{1}{2}$ lb of venison?

Sarah's new cookery book has different instructions.

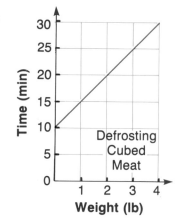

1 The graph shows the total defrost time for kidney.

(a) Copy and complete.

Weight in lb (W)	1	2	3	4
Time in min (T)	14			

Each extra pound needs 8 more minutes.

Formula: Time = 8 × Weight + 6

or $T = 8W + 6$

(b) Check that the formula works for 5 lb of kidney.
(c) What is the total defrost time for 9 lb of kidney?

2 For each of these graphs
- make a table
- write a formula
- check your formula.

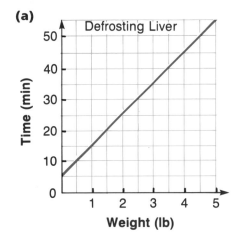

3 How long must Sarah allow for defrosting

(a) 8 lb of cubed meat
(b) 9 lb of liver
(c) 12 lb of chicken drumsticks
(d) $6\frac{1}{2}$ lb of cubed meat?

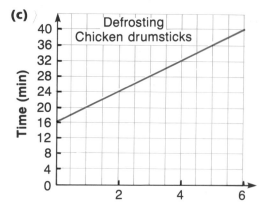

4 For each of the four graphs above

(a) give the point where the graph cuts the vertical axis
(b) explain any connection between your formula and this point.

Ask your teacher what to do next.

1 Use acute, obtuse, right, straight, or reflex to describe each of the angles marked in the picture.

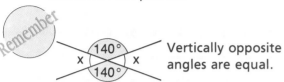

Remember

Vertically opposite angles are equal.

2 Name pairs of angles in the picture which are vertically opposite.

3 Find the value of each of the marked angles, *p* to *y*.

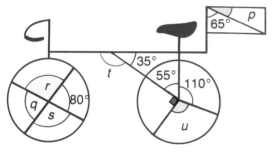

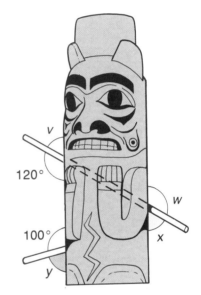

Remember

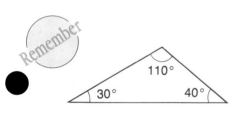

The sum of the angles in a triangle is 180°.

4 Copy each diagram and fill in the size of each missing angle.

(a)　　　　**(b)**　　　　**(c)**　　　　**(d)**

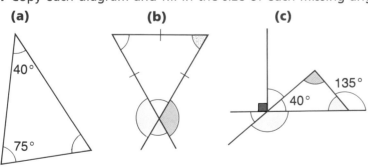

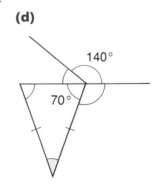

5 In this shape $\hat{DEF}$ is 50°. Write the name and size of each of the other angles.

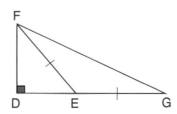

Challenge

Angles and parallel lines

This diagram shows a tiling of congruent parallelograms.

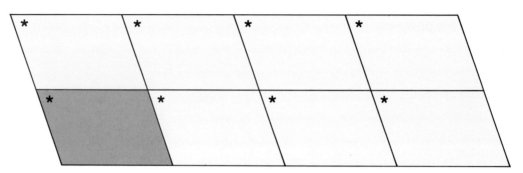

1 Trace the red parallelogram. Use your tracing to check that the * angles are equal.

Angles in the **same position** in each parallelogram are called **corresponding angles.**

2 Do Workbook page 7.

3 Copy each of the following diagrams and **without measuring**, find the values of angles *a* to *n*.

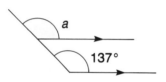

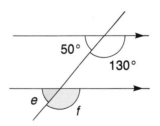

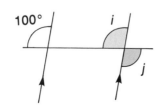

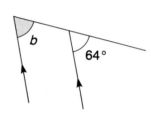

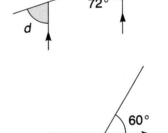

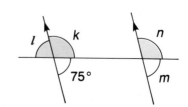

Here is another tiling of congruent parallelograms.

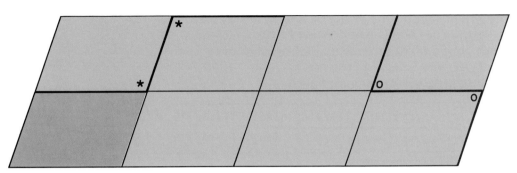

1 Trace the red parallelogram.
  Use your tracing to check that **each pair** of marked angles
  is equal.

  Pairs of angles like these are called **alternate angles**.

2 **Do Workbook page 8.**

3 Copy each of the following diagrams and **without
  measuring**, find the values of angles *m* to *z*.

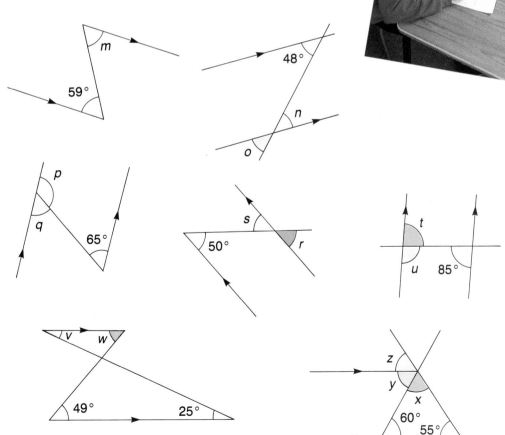

Ask your teacher what to do next.

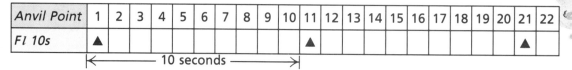

Each lighthouse has a different flash pattern.

Don, a navigator cadet on the *Seaforth,* checks his chart for Anvil Point lighthouse. The code is **Anvil Point *Fl* 10s** which means "during each period of 10 seconds the light flashes once".

This is the **code strip** for Anvil Point lighthouse.

Anvil Point	1	2	3	4	5	6	7	8	9	10	11	12	13	14	15	16	17	18	19	20	21	22
Fl 10s	▲										▲										▲	

← ——— 10 seconds ——— →

Later he checks the code for **Beachy Head** which is ***Fl*(2) 20s.** This means "during each period of 20 seconds the light flashes twice". There is one second between the flashes.

This is the **code strip** for Beachy Head lighthouse.

Beachy Head	1	2	3	4	5	6	7	8	9	10	11	12	13	14	15	16	17	18	19	20	21	22	23	24	25	26	27	2
Fl(2) 20s	▲		▲																		▲		▲					

← ——————— 20 seconds ——————— →

1 During the voyage the *Seaforth* passes other lighthouses.
Draw code strips for each of these.
- Mull of Galloway *Fl* 20s   • Bishop Rock *Fl*(2) 15s

2 Write the code for each of these lighthouses.

Lizard Point	1			4			7			10			13			16	
	▲			▲			▲			▲			▲			▲	

Portland Bill	1		3		5		7														21		23		25		27	
	▲		▲		▲		▲														▲		▲		▲		▲	

3 As the *Seaforth* enters the Bristol Channel it passes between Lundy Island and the mainland. Don times the flashes of the four lighthouses.

Match each set of times to a lighthouse.

> *Set A* 1,3,   21,23,   41,43,....
>
> *Set B* 1,3,5,7,9,11,   16,18,.....
>
> *Set C* 1,3,5,   11,13,15,   21,.....
>
> *Set D* 1,   6,   11,   16,   21,....

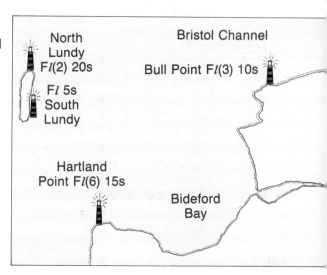

4 Don sees the two lights on Lundy
Island flash at exactly the same time.
How long will it be before he sees
them do so again?

5 As the *Seaforth* sails up the Firth of
Clyde the main lighthouses Don sees
are:

Pair P • **Turnberry Point** *Fl* 12s and
 Ailsa Craig *Fl*(6) 30s

Pair Q • **Pladda** *Fl*(3) 30s and
 Holy Island *Fl*(2) 20s

Pair R • **Little Cumbrae** *Fl* 3s and
 Toward Point *Fl* 10s.

For **each pair** of lighthouses use code
strips to investigate the following:

(a) • If the lights flash together once,
 will it happen again?
 If so, how long will this take?
 • Could they flash together again
 with a different length of time
 between flashes?
(b) Could there be a situation when
 the lights never flash together?
 Explain.

Ask your teacher what to do next.

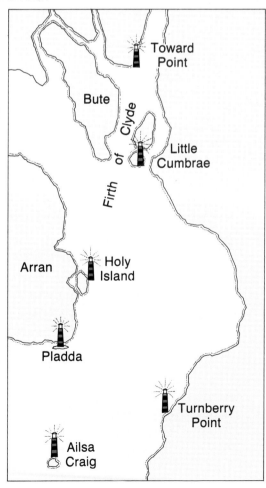

You can count or measure statistical data.
You count the number of students in a school.
You measure the heights of students.

Remember

1 Staff at Wellworthy Hospital collected the following data. Which of these did they
  • count   • measure?
  (a) the number of people in the hospital
  (b) the temperatures of patients
  (c) the weights of babies
  (d) the number of empty beds in the hospital.

You need Workbook page 9.

2 Dr Kildee works in casualty. Every day she records the number of patients treated. Here is her data for June.

| 42 | 56 | 39 | 68 | 62 | 46 | 78 | 54 | 84 | 69 | 54 | 65 | 66 | 56 | 34 |
| 74 | 87 | 59 | 96 | 43 | 67 | 77 | 94 | 62 | 72 | 49 | 85 | 74 | 28 | 65 |

(a) Find the range.
(b) Copy and complete the frequency table.
(c) Complete **Graph 1**.

Class interval	Tally marks	Frequency
20–29		
30–		

3 Rose works in the maternity unit. She records the number of babies born each day in September.

| 23 | 34 | 42 | 12 | 8 | 32 | 22 | 13 | 43 | 24 | 11 | 39 | 35 | 27 | 41 |
| 38 | 9 | 35 | 32 | 18 | 21 | 34 | 40 | 23 | 47 | 23 | 44 | 18 | 17 | 33 |

(a) Find the range.
(b) Choose suitable class intervals and draw a frequency table.
(c) Complete **Graph 2**.

4 Mike works in the hospital kitchen.
He has to know the ages of the children in Ward C and the number of meals needed for each ward.
For each set of data   • find the range
  • draw a frequency table
  • complete the graph.

(a) Data for Graph 3
Age of children in Ward C

1	10	12	6	7	7
11	4	4	6	12	9
5	7	4	5	15	8
3	13	5	8	7	12

(b) Data for Graph 4
Number of meal trays per ward

27	32	21	61	18	31
9	22	28	52	34	46
34	43	53	18	22	6
43	17	28	23	35	47

You need Workbook page 10.

5 Peter measures the time it takes for patients to be X-rayed. Here are his results in minutes.

10·5	12·2	16·8	23·4	26·8	34·1
11·8	22·6	34·6	12·7	23·4	21·6
15·8	26·9	23·0	37·5	24·8	23·0
16·7	17·1	17·9	33·9	24·6	19·5

(a) What is the range?
(b) Copy and complete the frequency table.
(c) Complete **Graph 1**.

Class interval	Tally marks	Frequency
10·0 – under 15·0		
15·0 – under 20·0		
20·0 – under 25·0		
25·0 – under 30·0		
30·0 – under 35·0		
35·0 – under 40·0		

For data which is measured you need class intervals like these.

6 Rose records the weights of new born babies. Here is her data for one day.

Weight of each baby in kilograms

3.4 3.5 4.1 2.0 3.3 3.9 3.2 3.1 2.3 2.6 3.6 3.3 2.6 3.6
3.8 1.7 2.7 3.2 4.1 3.5 3.9 3.0 2.5 3.6 4.4 3.6 4.2 2.7

(a) Copy and complete the frequency table.

Class interval	Tally marks	Frequency
1·5 – under 2·0		
2·0 – under 2·5		

(b) Complete **graph 2**.

7 Dr Kumar is collecting data for a research project. For each set of data
- find the range
- choose suitable class intervals and draw a frequency table
- complete the graph.

(a) **Data for Graph 3**
Exercise time in minutes

5·4	5·7	2·9	5·7	7·6	5·7
7·7	4·8	3·9	3·6	2·8	5·3
5·6	5·5	8·9	4·9	6·9	6·6
3·7	4·8	6·2	5·1	4·5	4·1

(b) **Data for Graph 4**
Fluid intake in litres

1·35	1·25	0·50	1·80	1·95	1·65
0·75	0·80	0·65	1·55	1·65	2·25
0·90	0·95	1·35	1·65	1·90	2·10
2·60	0·40	1·45	0·45	1·20	1·75

Is there a connection?

1 Ten boys recorded their
ages and heights.
The results are shown
on this **scatter graph**.

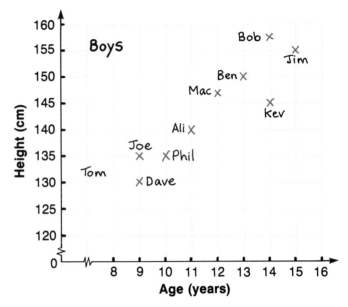

(a) How many boys are aged 9 years?
(b) How many boys are 140 cm tall?
(c) Who is the tallest?
(d) Who is the youngest?
(e) How old is Mac?
(f) How tall is the oldest boy?
(g) What can you say about Kev?
(h) What is the difference in height
between the 14 year old boys?

You need Workbook page 11.

2 Ten girls recorded their ages and heights. Use the
information to complete **scatter graph A.**

Name	Beth	Ann	Jill	Dot	Julie	Katy	Babs	Carla	Una	Mo
Age (years)	12	10	8	14	12	9	11	8	15	13
Height (cm)	140	135	125	155	155	135	145	135	160	150

3 (a) Find the height and the shoe size
of 10 boys or girls in your class.
Record the information in a table.
(b) Show this information on **scatter
graph B.**

Name		
Shoe size		
Height (cm)		

4 Find the height and weight in kilograms
of 10 boys or girls in your class.
Show this information on **scatter graph C.**

5 The table shows the number of days absent
and the Maths marks of 10 students.

Name	Kim	Eva	Sky	Khan	Ruby	Brian	Meg	Ali	Jarbir	Kurt
Days absent	4	20	4	2	18	16	12	20	22	9
Mark	20	8	18	17	9	12	13	5	4	16

Show this information on **scatter graph D.**

Your class is planning an activity weekend. Each student must be involved in **four** different activities, two indoor and two outdoor.

Work in a group.

1 Make a list of activities which you and your friends would enjoy on an activity weekend. Split the activities into two types:
  • indoor activities
  • outdoor activities
Select 5 of each type to offer on your weekend.

2 You need to know
  • each student's choice of activities
  • which activities are the most popular
  • whether the same activities are equally popular with boys and girls.
Make up a suitable survey sheet to collect the information you need.

3 Carry out a survey of your class using your survey sheet.

4 Organise the data from your survey and use it to draw at **least** two graphs.

5 Write about what you found in your survey.

Ask your teacher what to do next.

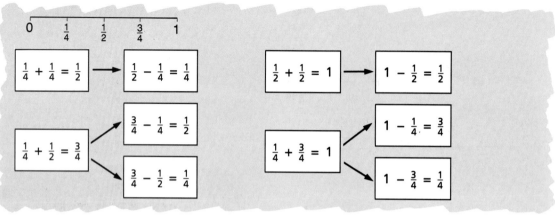

1 (a) Huw was holding a summer party. He bought a $\frac{1}{4}$ litre
and a $\frac{1}{2}$ litre carton of pineapple juice.
How much juice did he buy?

(b) He filled two $\frac{1}{4}$ litre glasses. How much pineapple juice was left?

2 Find **(a)** $1\frac{1}{4} + 1\frac{1}{4}$ **(b)** $2\frac{1}{2} + 2\frac{1}{2}$ **(c)** $2\frac{1}{2} + 1\frac{1}{4}$ **(d)** $5\frac{3}{4} + 2\frac{1}{4}$ **(e)** $4\frac{3}{4} - 1\frac{1}{2}$

(f) $2\frac{1}{2} - \frac{1}{4}$ **(g)** $4 - 1\frac{1}{2}$ **(h)** $2\frac{3}{4} - 1\frac{1}{4}$ **(i)** $6 - 4\frac{1}{4}$ **(j)** $3 - 2\frac{3}{4}$

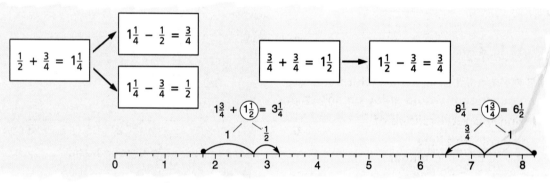

3 Find **(a)** $1\frac{1}{4} + \frac{1}{2}$ **(b)** $2\frac{1}{4} + 1\frac{3}{4}$ **(c)** $3\frac{1}{2} + \frac{3}{4}$ **(d)** $2\frac{3}{4} + 2\frac{3}{4}$ **(e)** $2\frac{3}{4} - \frac{1}{2}$

(f) $5\frac{3}{4} - 2\frac{1}{4}$ **(g)** $4 - 1\frac{3}{4}$ **(h)** $5\frac{1}{4} - 2\frac{1}{2}$ **(i)** $4\frac{1}{4} - 2\frac{3}{4}$ **(j)** $6\frac{1}{2} - 1\frac{3}{4}$

4 Huw bought a $\frac{3}{4}$ litre and a $1\frac{1}{2}$ litre carton of tomato juice.
How much tomato juice did he buy altogether?

The graph shows the other drinks
Huw bought for his party.

Huw's party drinks

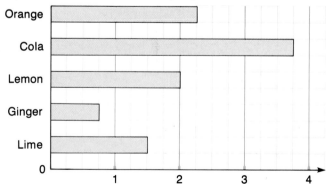

What was the total volume
of cola and lime?

$3\frac{3}{4} + 1\frac{1}{2}$
$= 4\frac{3}{4} + \frac{1}{2}$
$= 4\frac{3}{4} + \frac{2}{4}$
$= 4\frac{5}{4}$
$= 5\frac{1}{4}$ litres

$\frac{5}{4} = 1\frac{1}{4}$
4 and $1\frac{1}{4} = 5\frac{1}{4}$

5 Find the total volume of
(a) orange and ginger **(b)** lime and orange
(c) ginger and cola **(d)** cola, lemon and lime **(e)** ginger, lime and orange.

$\frac{1}{2} = \frac{2}{4} = \frac{4}{8}$

$\frac{1}{2} \xrightarrow{\times 2} \frac{2}{4} \xrightarrow{\times 2}$ $\frac{1}{2} \xrightarrow{\times 4} \frac{4}{8} \xrightarrow{\times 4}$

	1 whole
	2 halves
	4 quarters
	8 eighths

Jackie's party drinks

Huw helped organise drinks for Jackie's birthday party. They bought these drinks.

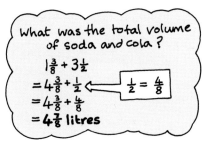

What was the total volume of soda and cola?

$1\frac{3}{8} + 3\frac{1}{2}$

$= 4\frac{3}{8} + \frac{1}{2}$ $\frac{1}{2} = \frac{4}{8}$

$= 4\frac{3}{8} + \frac{4}{8}$

$= 4\frac{7}{8}$ litres

Bar chart titled "Jackie's party drinks" with drinks Ginger, Lemon, Orange, Cola, Tonic, Lime, Soda. Volume in litres, axis 0 to 4.

Volume in litres

1 Find the total volume of

(a) cola and lime
(b) lemon and ginger
(c) soda and lime
(d) orange and soda
(e) tonic and ginger.

2 Find **(a)** $2\frac{5}{8} + 3\frac{1}{2}$ **(b)** $1\frac{3}{8} + 2\frac{3}{4}$ **(c)** $\frac{7}{8} + 2\frac{1}{4}$ **(d)** $2\frac{3}{4} + 2\frac{5}{8}$ **(e)** $\frac{7}{8} + 1\frac{3}{8}$

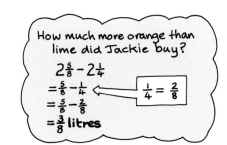

How much more orange than lime did Jackie buy?

$2\frac{5}{8} - 2\frac{1}{4}$

$= \frac{5}{8} - \frac{1}{4}$ $\frac{1}{4} = \frac{2}{8}$

$= \frac{5}{8} - \frac{2}{8}$

$= \frac{3}{8}$ litres

3 How much more did Jackie buy of

(a) cola than soda
(b) cola than lemon
(c) ginger than soda
(d) orange than soda
(e) cola than lime?

4 Find **(a)** $3\frac{7}{8} - 1\frac{1}{4}$ **(b)** $2\frac{5}{8} - 1\frac{3}{8}$ **(c)** $4\frac{7}{8} - 2\frac{1}{2}$

(d) $2\frac{7}{8} - 1\frac{3}{4}$ **(e)** $3\frac{3}{4} - 1\frac{5}{8}$

5 Find how much more lemon Jackie bought than

(a) lime **(b)** orange **(c)** tonic.

Challenge

Pignatelli's Pizzeria serves small, medium and large pizzas.

In week one they sold 600 pizzas.

$\frac{9}{20}$ were small pizzas.

How many small pizzas is this?

$\frac{1}{20}$ of 600 = 600 ÷ 20 = 30

So $\frac{9}{20}$ of 600 = 30 × 9 = 270

270 were small pizzas.

1 In week one $\frac{2}{5}$ of the pizzas sold were large pizzas.
  (a) How many large pizzas did they sell?
  (b) How many medium pizzas did they sell?

2 The pizzeria sells four types of pizza
   Giovanna keeps a record of the fraction of each type sold.

Type of pizza	Napoli	Funghi	Cipolle	Pepperone
Fraction sold	$\frac{1}{5}$	$\frac{3}{10}$	$\frac{3}{20}$	$\frac{7}{20}$

  (a) In week two they sold 800 pizzas. How many of each type did they sell?
  (b) In week three they sold 760 pizzas. How many of each type did they sell?

3 There were 64 customers in the pizzeria. $\frac{3}{4}$ ordered garlic bread.
   How many customers ordered garlic bread?

   Pignatelli's recipes contain these ingredients.

	Cheese	Tomato	Flour
Large Pizza	$\frac{3}{10}$ kg	$\frac{4}{5}$ tin	$\frac{3}{8}$ kg
Medium Pizza	$\frac{1}{10}$ kg	$\frac{3}{5}$ tin	$\frac{3}{20}$ kg

How much flour is needed for 10 large pizzas?

$10 \times \frac{3}{8} = \frac{30}{8}$

$\qquad = 3\frac{6}{8}$

$\qquad = 3\frac{3}{4}$

$3\frac{3}{4}$ kg of flour is needed.

4 How much of **each** ingredient is needed for
  (a) 5 large pizzas   **(b)** 8 large pizzas
  (c) 11 large pizzas   **(d)** 10 medium pizzas
  (e) 7 medium pizzas   **(f)** 12 medium pizzas
  (g) 20 large pizzas   **(h)** 16 medium pizzas?

The Spend 'n' Save! store has a new line of better value products.

Spend 'n' Save! Cola contains $\frac{1}{3}$ more.

$\frac{1}{3}$ of 330 ml = 330 ÷ 3 = 110 ml

So the new Cola contains 330 + 110 = **440 ml**

1 Find the size of each of these new Spend 'n' Save! products.

(a) Old size 500g — Krunchy Fruit Cereal $\frac{1}{4}$ more

(b) Old size 200 sheets — TOILET ROLLS $\frac{1}{5}$ more

CHOCKY STICKS $\frac{1}{3}$ more

(c) Old size 400g — CHEDDAR CHUNK $\frac{1}{4}$ EXTRA

TOOTHPASTE $\frac{1}{4}$ MORE — Old size 120 ml

(d) Old pack 60 sticks **(e)**

Janice, the manager, has reduced the price of some of Spend 'n' Save! products to attract more customers.

HAM £9.50

REDUCE BY $\frac{2}{5}$

Spend 'n' Save! ham has been reduced by $\frac{2}{5}$.

$\frac{1}{5}$ of £9·50 = £1·90

$\frac{2}{5}$ of £9·50 = £1·90 × 2 = £3·80

So the ham costs £9·50 − £3·80 = **£5·70**

2 Find the new price of each of these Spend 'n' Save! products.

(a) LEMONADE 20p — REDUCE BY $\frac{1}{5}$

(b) Mince Pies £1·20p — $\frac{2}{3}$ OFF

(c) 88p — REDUCE BY $\frac{3}{4}$

(d) BUTTER 68p — $\frac{1}{4}$ OFF

(e) BEANS 30p — $\frac{3}{10}$ OFF

3 Janice has also changed some of the meat prices. Find the new price of each item in the table.

Item	Old Price	Change
Frozen Chicken	£4·50	increase by $\frac{1}{5}$
Steak Pie	£2·82	reduce by $\frac{1}{3}$
Bacon	£12·50	$\frac{1}{10}$ less
Leg of Pork	£8·64	$\frac{1}{4}$ more

4 Kenneth is buying soap powder in Spend 'n' Save!.
Which box is the better value?
Give a reason for your answer.

Old size 750g — SUDSO $\frac{1}{3}$ MORE POWDER £1·65

750g SUDSO £1·65 PRICE REDUCED BY $\frac{1}{5}$

Challenge

Ask your teacher what to do next.

You need Workbook page 12.
Secret Agent Kooper is gathering information about an enemy agent. Solve the problems below to help her.

1 Use the clues in the notebook below, together with the list in the Workbook, to find the country in which the agent is hiding.

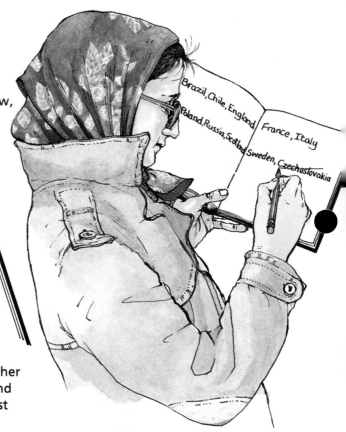

- The name of the country has less than 9 letters.
- The name of the country does not have exactly 6 letters.
- The country's name does not start with S.
- The enemy agent is not in South America.
- The capital city of the country is not Rome.

2 Use the clues in the telegram, together with the list in the Workbook, to find the town in which the agent was last spotted.

NAME OF TOWN HAS MORE THAN 4 LETTERS STOP HAS AN EVEN NUMBER OF LETTERS STOP DOES NOT CONTAIN 3 LETTER 'A's STOP DOES NOT START OR END WITH AN 'H' STOP

3 Kooper knows that the agent's hideout is between 30 and 60 miles from London. Use the clues to pinpoint the distance.

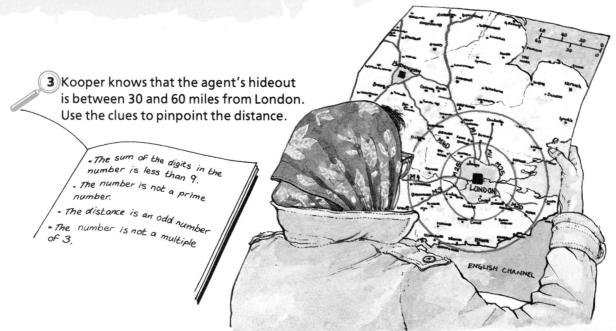

- The sum of the digits in the number is less than 9.
- The number is not a prime number.
- The distance is an odd number.
- The number is not a multiple of 3.

4 Use the clues in this secret file to find the agent's age in years.

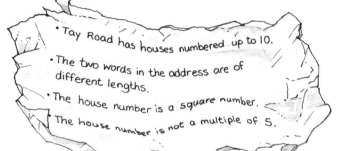

```
                                        TOP SECRET
AGENT NAME : UNKNOWN

AGENT AGE :  ■ AN EVEN NUMBER BETWEEN 21 AND 51
             ■ THE SUM OF THE DIGITS IS BETWEEN
               7 AND 10
             ■ THE NUMBER IS NOT A MULTIPLE OF 4
```

5 Kooper has entered some information in the table in the Workbook about the agent's hideout.
Use the clues to find the address.
Put crosses in the table as you eliminate each address.
The crosses for the first clue have been entered for you.

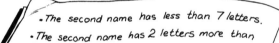

- Tay Road has houses numbered up to 10.
- The two words in the address are of different lengths.
- The house number is a square number.
- The house number is not a multiple of 5.

6 Kooper discovers that the agent's first name is either Amanda, Henrietta, John, Maria or Tom. The agent's second name is either Adams, Bond, Stanhope or Trapper. Use these clues and the table in the Workbook to find the agent's full name.

- The second name has less than 7 letters.
- The second name has 2 letters more than the first.

7 The enemy agent has four assistants called Abel, Bluff, Conman and Dunn.
Study the drawings and use the clues to match the correct name to each one.

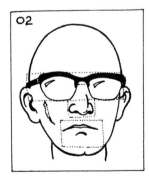

- Abel does not have a scar
- Bluff has never had a beard
- Dunn is not bald
- Conman is always clean-shaven and always wears glasses.

You need Workbook page 13.

1 Chris makes rugs 5 feet by 3 feet from these carpet cut-offs.

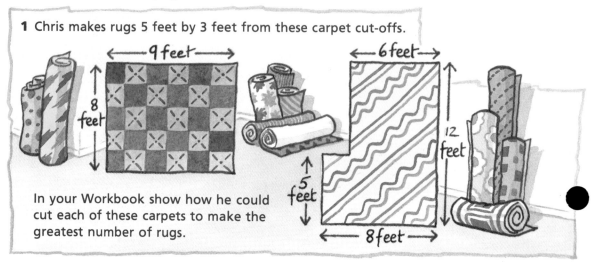

In your Workbook show how he could cut each of these carpets to make the greatest number of rugs.

2 Chris employs 3 carpet fitters: Callum, Fiona and Shamsad.
They each work from 9 am to 5 pm with 1 hour off for lunch.
They have to do these jobs on Wednesday.

Chris makes up daily work schedules for his fitters. He started Wednesday's schedule by giving Job 2 to Callum.

(a) Complete Wednesday's schedule in your Workbook.

(b) Here is Thursday's work. Complete the schedule in your Workbook.

Strangehill School has a maths club every Tuesday after school. The members play games and solve problems and puzzles.

The fifth degree

Work with a partner. You need a protractor.

How to play
- Player 1 draws and measures an angle without letting Player 2 see the size.
- Player 2 has 5 chances to guess the size of the angle.
- After each guess Player 1 says whether Player 2's guess is correct (if within 5°) or too big or too small.
- The winner is the first player to score more than 12.

How to score

Number of guesses needed	1	2	3	4	5
Points	7	4	3	2	1

Chicken feed

Aunt Mabel keeps chickens. They are kept in a rectangular run with a perimeter of 68 metres. The length of the run is 8 metres more than its width. How long is the run?

Digit deduction

(a) 5 is divided by a single-digit number to give

$$0.7142857$$

Find the number.

(b) One single-digit number is divided by another single-digit number. Find the numbers which give

$$0.7777777$$
$$1.2857124$$

What's for starters?

For this flow chart find the starting number which will give the answer

(a) 12

(b) 210

(c) 91.91

(d) 11.31

(e) 16

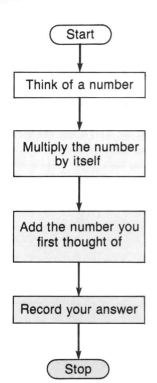

Start

→ Think of a number

→ Multiply the number by itself

→ Add the number you first thought of

→ Record your answer

→ Stop

1 This scroll contains two problems from the test you must take before you can join the King of Drakonia's Imperial Guard. Solve the problems and become a Guard!

In each diagram the sum of the numbers on each line should be the same.

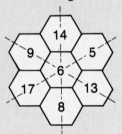

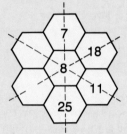

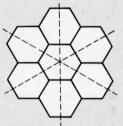

In this diagram the sum is 28.

(a) Trace and complete this diagram.

(b) Trace and complete this diagram using each of the numbers 11 to 17 once only.

2 These four soldiers are from the Imperial Guard.

(a) Use the clues to name the soldiers from left to right.

- Thumpah is taller than Klouta.
- Nuttah is shorter than Klouta.
- Basha is the shortest of them all.

(b) Each of these four soldiers must choose a **different** weapon from the store – a short sword, a long sword, an axe or a bow. List all the different ways in which they could choose their weapons.

3 Each guard welds five pieces of metal like these together to make a square shield. Show how the pieces fit.

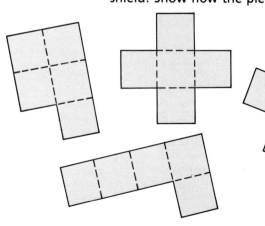

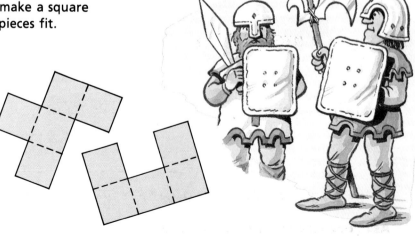

4 These are the four types of warriors in
the Drakonian army.

LONG SWORDSMAN   SHORT SWORDSMAN   AXEMAN   BOWMAN

In battles the warriors march in the Drakonian Square
Formation which has 4 rows each having 4 warriors. No row,
column or diagonal has more than one warrior of the same
type.
Draw a diagram to show how the warriors could arrange
themselves in Square Formation.

5 The Drakonian Army is well known for its achievements in
battle against the evil Dark Prince of Zandor.

The land of Zandor is circular. The evil Prince wants to build
four **straight** roads to divide his land into ten parts of
various sizes, one part for each of the ten tribes of Zandor.
Draw a diagram to show how he could build the roads.

6 When the victorious Drakonian Army returned from Zarok,
the capital of Zandor, it had to march to **one** of the wells of
Pyrus to collect water supplies. This map shows a possible
route.

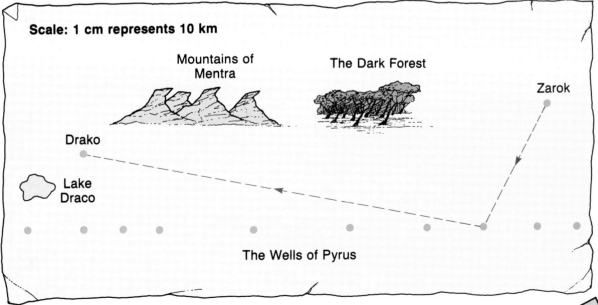

Scale: 1 cm represents 10 km

Mountains of
Mentra

The Dark Forest

Zarok

Drako

Lake
Draco

The Wells of Pyrus

Use the map on **Workbook page 12** to find the shortest possible route.

Ask your teacher what to do next.

1 kilogram = 1000 g

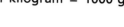

10 of these weigh 1 kg

$100 \text{ g} = \frac{1}{10} \text{ kg} = 0.1 \text{ kg}$

$900 \text{ g} = \frac{9}{10} \text{ kg} = 0.9 \text{ kg}$

100 of these weigh 1 kg

$10 \text{ g} = \frac{1}{100} \text{ kg} = 0.01 \text{ kg}$

$90 \text{ g} = \frac{9}{100} \text{ kg} = 0.09 \text{ kg}$

1000 of these weigh 1 kg

$1 \text{ g} = \frac{1}{1000} \text{ kg} = 0.001 \text{ kg}$

$9 \text{ g} = \frac{9}{1000} \text{ kg} = 0.009 \text{ kg}$

1 Write each of these weights as
- a fraction of a kilogram
- a decimal fraction of a kilogram.

(a) 300 g **(b)** 30 g **(c)** 3 g **(d)** 70 g **(e)** 700 g **(f)** 7 g

The total weight is $\frac{2}{10} + \frac{1}{100} + \frac{3}{1000}$ kg

As a decimal fraction it is 0.213 kg.

2 In the same way write each total weight
- using fractions
- as a decimal fraction of a kilogram.

(a)  **(b)**  **(c)**

3 The electronic scale shows weights in kilograms.
What weight is displayed when an object weighs
(a) 3765 g **(b)** 2047 g **(c)** 1005 g **(d)** 686 g
(e) 409 g **(f)** 83 g **(g)** 40 g **(h)** 5 g ?

4 Do Workbook page 14.

5 For each set, list the three weights in order starting with the
lightest.
(a) 1.09 kg, 1.1 kg, 1.089 kg **(b)** 0.4 kg, 410 g, 0.39 kg,
(c) $2\frac{3}{10}$ kg, 2200 g, 2.487 kg **(d)** 5050 g, $5\frac{1}{2}$ kg, 5.250 kg

6 Write **your** weight
(a) to the nearest kilogram
(b) as a decimal fraction of a tonne.

7 Write the weight of each vehicle as a
decimal fraction of a tonne.
(a) car **(b)** truck **(c)** van
900 kg 12 708 kg 1017 kg

Class 3 at Strangehill visited their local airport.

1 (a) Find the total weight of luggage carried
by each of these passengers.

(b) Each passenger is allowed to take 33 kg of luggage.
Whose luggage is too heavy and by how much?
(c) For each passenger find the difference in weight between
their two pieces of luggage.

2 Class 3 watched Andy operating the
weighbridge. He weighed goods
vehicles when they entered the airport
and again when they left. He kept this
record.

(a) Find the weight of goods
unloaded from each vehicle.
(b) Find the total weight of goods
loaded on flight
  • LN 486 • PQ 338 • MK 047.

Vehicle	Weight of vehicle		Goods for flight no.
	entering airport	leaving airport	
Smith's	2·867 tonnes	1·435 tonnes	LN 486
C and Q	1·659 tonnes	0·253 tonnes	PQ 338
Tariq's	2·742 tonnes	1·482 tonnes	LN 486
Fastline	3·358 tonnes	1·633 tonnes	MK 047
Scotpost	2·139 tonnes	0·862 tonnes	MK 047
Overnight	3·026 tonnes	1·489 tonnes	PQ 338

3 Some of Class 3 played an arcade game, *Emergency Stop*. The
player drives a car and a dog runs in front of it. The player's
reaction time is measured.

EMERGENCY STOP-REACTION TIME IN SECONDS			
BOYS		GIRLS	
MUMTAZ	0·343	LISA	0·428
JOE	0·336	CAROLE	0·389
WES	0·408	SHAREEN	0·355
BILLY	0·361	LINDA	0·392

(a) Who was the quickest boy?
(b) How much quicker was he than each of
the other boys?
(c) Who was the quickest girl?
(d) How much quicker was she than each of
the other girls?

4 Find the difference between the **average**
reaction times of the boys and of the girls.

A Touch of Class

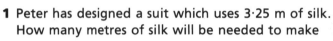

A Touch of Class designs and
produces clothes for women.

1 Peter has designed a suit which uses 3·25 m of silk.
How many metres of silk will be needed to make

(a) 3 suits   (b) 5 suits   (c) 9 suits?

2 Sarah supplies the
material needed to
make the clothes.
Calculate the total
length of material
required for each style
of dress ordered by La
Belle Boutique.

ORDER La Belle Boutique			
Dress Style	Length per dress in metres	Number Ordered	Total length required in metres
A	2·6	6	
B	3·75	5	
C	4·85	9	
D	4·18	8	
E	6·05	3	
F	7·25	7	

To multiply by 40 multiply by 4, then by 10 **or** multiply by 10, then by 4
To multiply by 700 multiply by 7, then by 100 **or** multiply by 100, then by 7

3 Suters has placed a big
order for six different
styles of suits. What
total length of material
is required for each
style?

ORDER Suters			
Suit Style	Length per suit in metres	Number Ordered	Total length required in metres
G	5·7	60	
H	4·83	500	
I	3·75	70	
J	4·55	90	
K	6·85	400	
L	2·95	800	

4 Jon designed a wedding dress which uses 5·8 m of silk,
0·55 m of lace and 1·33 m of ribbon. Find the length of
each material needed to make 200 wedding dresses.

5 Theresa is working on designs in denim.
The dress will use 2·8 m and the suit will use 3·6 m.
How much denim will be left from a 500 m roll after
80 dresses and 50 suits are made?

6 A ballroom dress is trimmed using 4·25 m of sequins.
  (a) There are 300 sequins to the metre. How many
     sequins are there on **each** dress?
  (b) How many sequins are used to trim 50 dresses?

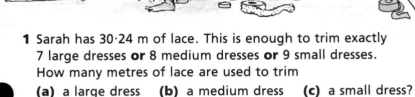

1 Sarah has 30·24 m of lace. This is enough to trim exactly
7 large dresses **or** 8 medium dresses **or** 9 small dresses.
How many metres of lace are used to trim

   (a) a large dress   **(b)** a medium dress   **(c)** a small dress?

2 Francesca's Frocks
ordered dresses
trimmed with coloured
braid. For each colour
of braid calculate the
length used per dress.

ORDER *Francesca's Frocks*

Braid colour	Length per dress in metres	Number Ordered	Total length of braid in metres
red		3	4·8
blue		4	6·04
green		5	9·30
yellow		8	17·2
pink		7	5·81

Remember

To divide by 40 divide by 4, then by 10 **or** divide by 10, then by 4
To divide by 700 divide by 7, then by 100 **or** divide by 100, then by 7

3 The Glass Slipper
ordered dresses
trimmed with ribbon.
For each style of dress,
calculate the length of
ribbon used per dress.

ORDER *The Glass Slipper*

Dress Style	Length per dress in metres	Number Ordered	Total length of ribbon in metres
M		20	8·4
N		30	12·3
O		50	34·55
P		400	124·4
Q		300	236·1
R		800	433·6

4 Copy and complete the stock control form for the length of silk used during August.

Stock Control *August* Material *Silk*			
Dress Style	Length per dress	Number of dresses made	Total length used
Wedding	7·93 m	6	
Shalwar		9	29·34 m
Cocktail		7	34·51 m
Kameez	8·45 m	4	
Ball gown	6·85 m		61·65 m

Challenge

The students and staff at Strangehill collect aluminium cans and paper for recycling. In September they collected 61·9 kg of paper and tied it up in 33 bundles. The average weight of a bundle of paper can be calculated like this:

Enter **61.9** Press **÷ 3 3 =** to give **1.8757575**

1·8757575 is between 1·8 and 1·9
It is nearer 1·9

The average weight of a bundle of paper is **1·9 kg to one decimal place.**

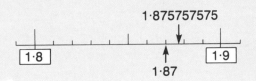

1·875757575

1·8 1·87 1·9

1 The students keep a record of the paper they collect. For each month in the students' table, calculate to one decimal place the average weight in kg of a bundle of paper.

Paper	Oct	Nov	Dec	Jan	Feb	Mar
Weight in kg	107·3	121·8	73·4	84·3	88·2	94·6
Number of bundles	50	56	34	39	41	46

1·8757575 is between 1·87 and 1·88
It is nearer 1·88

In September the average weight of a bundle of paper is **1·88 kg to two decimal places.**

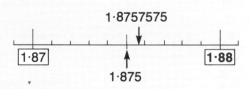

1·8757575

1·87 1·875 1·88

2 (a) For each month in the students' table, calculate to two decimal places the average weight in kg of a bag of aluminium cans.

Aluminium cans	Oct	Nov	Dec	Jan	Feb	Mar
Weight in kg	27·31	34·81	22·76	30·07	27·71	32·38
Number of bags	11	14	9	12	22	13

(b) In which month were smaller bags used? Explain.

3 (a) Strangehill receives £0·14 per kg for paper. Find, to the nearest penny, the amount of money raised each month from collecting paper.

(b) They receive £0·47 per kg for aluminium. Find, to the nearest penny, the amount of money raised each month from collecting cans.

Ask your teacher what to do next.

You can find factor pairs of 30 like this:

1 × 30 = 30   2 × 15 = 30   3 × 10 = 3   5 × 6 = 30

```
   30              30              30              30
  /  \            /  \            /  \            /  \
 1    30         2    15         3    10         5    6
```

The factors of 30 are 1, 2, 3, 5, 6, 10, 15, 30.

You can find factor pairs of 12 like this:

```
   12              12           12
  /  \            /  \         /  \
 1    12         2    6       3    4
```

The factors of 12 are 1, 2, 3, 4, 6, 12.

1, 2, 3 and 6 are factors of **both** 30 and 12.
They are called the **common factors** of 30 and 12.

1 For each pair of numbers
- find all the factors of each number
- list the common factors.

(a)

(b)

(c)

(d)

(e)

(f)

The multiples of 4 are 4, 8, 12, 16, 20, 24, 28 . . . .
The multiples of 6 are 6, 12, 18, 24, 30, 36, 42 . . . .

12 and 24 are multiples of **both** 4 and 6.
They are called **common multiples** of 4 and 6.

2 Find two common multiples for each pair of numbers.

(a)

(b)

(c)

(d)

(e)

(f)

3 For each set of numbers find
- the lowest common multiple
- the highest common factor.

(a)

(b)

(c)

Challenge

Ask your teacher what to do next.

Rahul has been designing brochures for Brushstrokes Art Gallery, using photographs of famous paintings.
He found the scale of "The Shrimp Girl" photograph like this:

Height of photograph = 6 cm
Height of painting = 60 cm

Height of photograph → ✕ 10 → Height of painting

Scale = 1:10

Check:

Width of photograph = 5 cm
Width of painting = 50 cm

Width of photograph → ✕ 10 → Width of painting

Scale = 1:10

W. Hogarth "The Shrimp Girl"
Height 60 cm   Width 50 cm

1 For each of the other photographs:
(a) use **one** of the dimensions to calculate the scale
(b) check that the other dimension gives the same scale.

J. Reynolds
"Portrait of Nelly O'Brien"
Height 125 cm   Width 100 cm

E. Munch
"The Scream"
Height 90 cm
Width 72 cm

J.B. Chardin
"Bouquet de Fleurs"
Height 42 cm
Width 36 cm

A. Renoir
"La Loge"
Height 84 cm
Width 60 cm

2 (a) Measure the height of "Self-portrait".
  (b) Calculate the scale.
  (c) Use the scale to calculate the width of the painting.

3 For each of the other photographs
  (a) calculate the scale
  (b) use the scale to calculate the missing dimension.

*V. van Gogh
"Self-portrait"
Height 65 cm
Width* ☐

*A. Ramsay "Portrait of the Artist's Wife"
Height* ☐ *Width 63 cm*

*E. Manet "En Bateau"
Height 96 cm   Width* ☐

*J-F. Millet "Wood Sawyers"
Height* ☐ *Width 99 cm*

*L. da Vinci "Mona Lisa"
Height* ☐ *Width 50 cm*

4 Do Workbook page 15.

**Speed:
Distance,
speed, time**

Many types of boat cross
the Channel between
England and France.

RAMSGATE

Scale **1 cm to 10 km**

DOVER
FOLKESTONE

DUNKIRK

CALAIS

BOULOGNE

1 Find the true distance
between Dover and
  (a) Boulogne
  (b) Calais
  (c) Dunkirk.

2 Find the average speed
of each boat in
kilometres per hour.

(a)

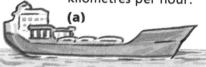

45 km in 3 hours

(b)

**Ramsgate to Calais
in 5 hours**

(c)

**Folkestone to Calais
in 4 hours**

(d)

**Dover to Dunkirk
in 10 hours**

3 Explain how to calculate the
average speed when you know
the distance and the time taken.

4 The car ferry *Cygnet* sails
at an average speed of 20
kilometres per hour. Copy
and complete the table:

Speed 20 km/h

Distance	5 km	10 km	20 km	40 km	60 km	80 km	100 km
Time	$\frac{1}{4}$ hour		1 hour				

5 How long should the *Cygnet* take to sail
  (a) 30 km   **(b)** 70 km   **(c)** 25 km   **(d)** 65 km   **(e)** 15 km   **(f)** 35 km?

6 How long should it take the *Cygnet* to sail from Dover to
  (a) Boulogne   **(b)** Calais   **(c)** Dunkirk?

7 Explain how to calculate the **time** taken when you know the
distance and the average speed.

8 (a) *Cygnet* leaves Dover for Dunkirk at 0745 hours.
    When should it arrive in Dunkirk?
  (b) The hovercraft *Juno* travelled at an average speed of
    40 kilometres per hour between Folkestone and Boulogne.
    It arrived in Boulogne at 1805 hours.
    When did it leave Folkestone?

1 The hovercraft *Cumbria* travels at an
average speed of 30 kilometres per
hour. Copy and complete the table.

Time	$\frac{1}{4}$ hour	$\frac{1}{2}$ hour	1 hour	2 hours	3 hours	4 hours
Distance	$7\frac{1}{2}$ km		30 km			

2 How far should the *Cumbria* travel in

(a) $3\frac{1}{2}$ hours   **(b)** $1\frac{1}{4}$ hours   **(c)** 45 min   **(d)** 2 hours 15 min   **(e)** $2\frac{3}{4}$ hours?

3 Between which two ports could *Cumbria* be travelling if the journey time is

(a) 2 hours   **(b)** 1 hour 30 min   **(c)** 2 hours 30 min?

4 Explain how to calculate the **distance** travelled when you know the average speed
and the time taken.

5 (a) *Sulvan* sails 75 km in 3 hours.
What is its average speed?

(b) *Seaforge* sails 100 km at 25 km/h.
How long does the journey take?

(c) *Polo* sails at 21 km/h for 4 hours.
What is the distance travelled?

(d) *Seamaiden* sails 90 km in 5 hours.
What is its average speed?

(e) *Riva* sails at 24 km/h for $2\frac{1}{2}$ hours
What is the distance travelled?

(f) *Sunrise* sails 90 km at 40 km/h.
How long does the journey take?

Ask your teacher what to do next.

Jack designs gift boxes for the company Boxing Clever.

1 On this gift box Jack has coloured opposite faces the same.
Each red face measures 3 cm by 2 cm.
What is the size of each

  (a) yellow face   **(b)** green face?

2 (a) Jack has sketched the different sized faces of the box
    before making a net.
    • Complete net (P) on **Workbook page 22.**
    • Cut it out and check that it folds to make the box.
  (b) Calculate the area of **each** face of the box.
  (c) Find the total area of all six faces.

> This is called the **total surface area.**

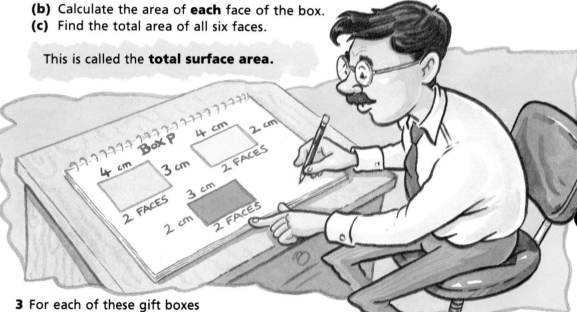

3 For each of these gift boxes
  • sketch the different sized faces
  • complete the net on **Workbook page 22**
  • cut out your net and check that it folds to make the box
  • calculate the total surface area.

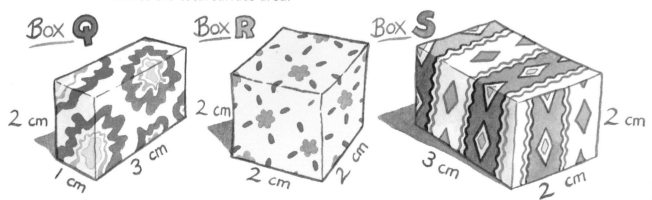

4 For each of these gift boxes
  • sketch the different sized faces
  • on 1 cm square dot paper draw a net
  • cut out your net and check that it folds to make the box
  • calculate the total surface area.

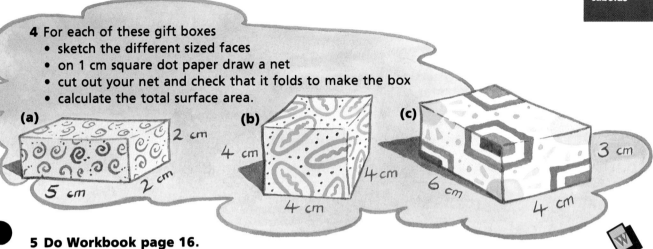

(a) 2 cm 5 cm 2 cm

(b) 4 cm 4 cm 4 cm

(c) 3 cm 6 cm 4 cm

5 Do Workbook page 16.

6 Here are nets of some gift boxes. For each box, which fruit
will be on the face opposite the apple?

(a)   **(b)**   **(c)**

7 Jack has designed some gift boxes for young children. The
pictures on the faces at vertex L are a car, a truck and a train.
List the pictures at vertex

(a) M   **(b)** N.

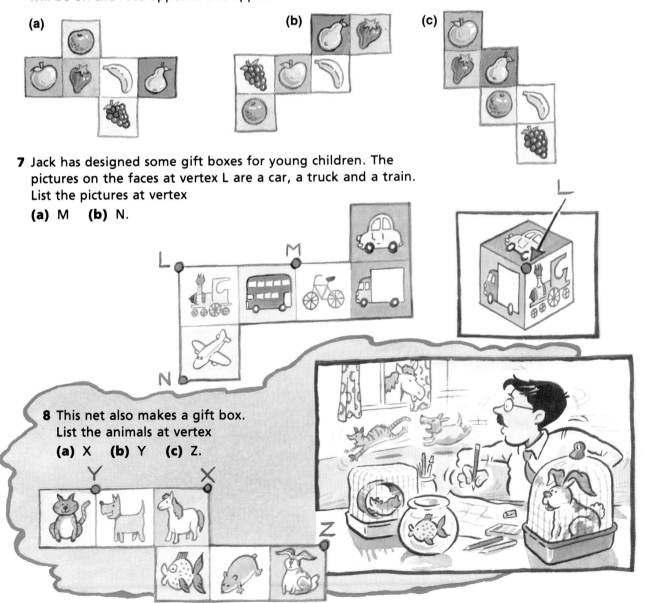

8 This net also makes a gift box.
List the animals at vertex
  (a) X   **(b)** Y   **(c)** Z.

Jack also designs novelty gift boxes.

1 (a) Name the shape of box ⓣ.

(b) Jack sketched the different sized faces of the box before making a net.
  • Complete net ⓣ on **Workbook page 22.**
  • Cut it out and check that it folds to make the box.

(c) Calculate the total surface area of the box.

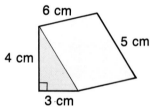

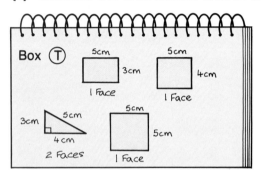

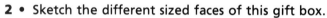

2 • Sketch the different sized faces of this gift box.
  • Draw its net on 1 cm square dot paper.
  • Cut out your net and check that it folds to make the box.
  • Calculate the total surface area of the box.

3 Repeat question **2** for each of these novelty gift boxes.

(a)

(b)

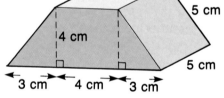

4 Make a novelty gift box for this model car.

length 8 cm
breadth 3 cm
height 2 cm

Challenge **5** Jack makes a novelty gift box from this net of a **tetrahedron.** Make an accurate model of the box.

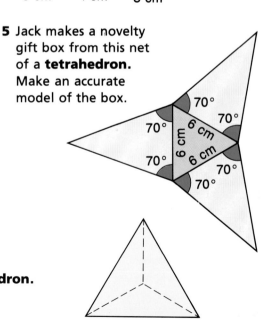

Challenge **6** This novelty gift box is a **regular tetrahedron.** The length of each of its edges is 7 cm. Make an accurate model of the box.

Sally makes different solids from cubes and sketches them.

1 Do Workbook page 17.

2 On isometric dot paper sketch each solid.

(a)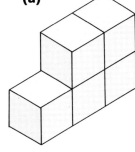

(b)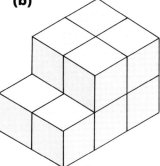

(c)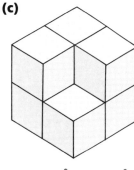

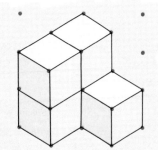

Here is Sally's sketch of what this solid looks like with the red cubes removed.

3 On isometric dot paper draw each of these solids after the red cubes have been removed.

(a)

(b)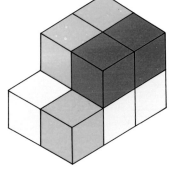

(c)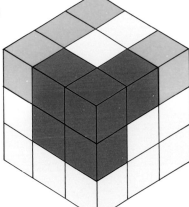

4 Look at the original solids in question **3**.
One cube is placed on top of each blue cube. Sketch the solids.

Challenge

5 How many different solids could Sally build using
  (a) 3 cubes   **(b)** 4 cubes?

Challenge

Ask your teacher what to do next.

You need A4 paper, a ruler and compasses.

1 This view of the satellite TV aerial shows a **parabola.** Follow these steps to draw a parabola.

Step 1

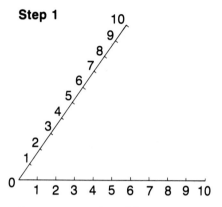

- Draw an angle with arms 10 cm long.
- Mark and number each centimetre along the arms.

Step 2

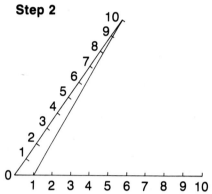

- Join point 1 on one arm to point 10 on the other.

Step 3

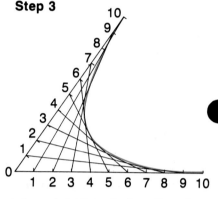

- Join points 2 to 9, 3 to 8 and so on.
- Draw the smooth curve shown by the red line.

The curve you have drawn is called the **envelope** of the lines. This envelope is in the shape of a **parabola.**

2 Follow these steps to draw another envelope.

Step 1

Step 2

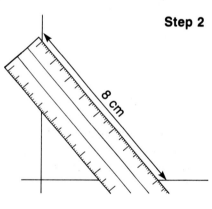

8 cm

Step 3

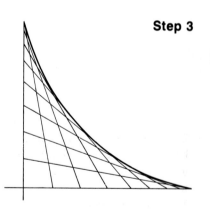

- Draw 2 lines each 16 cm long, which bisect each other at right angles.

- In the top right-hand quarter of the diagram, draw a line 8 cm long as shown.

- Draw more lines each 8 cm long.

Step 4: Repeat Step 3 in each of the other quarters of your diagram. Draw the envelope of your lines.

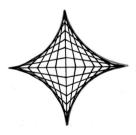

This envelope is called an **astroid.**

3 You can draw envelopes from **circles.** Follow these steps.

Step 1	Step 2	Step 3
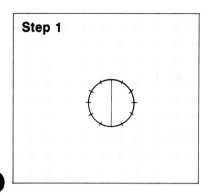	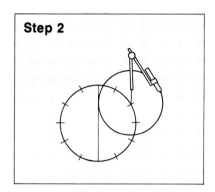	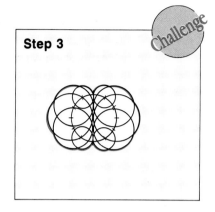

- Draw a circle, radius 5 cm, in the centre of a sheet of paper.
- Draw a diameter and mark 10 points on the circle, as shown.

- Using one of your points as the centre, draw another circle which **just touches** the diameter.

- Repeat Step 2 for each of your other points.
- Draw the envelope.

The envelope of these circles is called a **nephroid** (or kidney shape).

4 (a) Try this envelope. It is called a **limaçon** (or snail).

Step 1	Step 2	Step 3
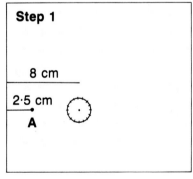	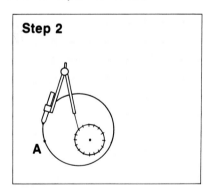	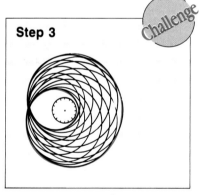

- Draw a circle, radius 3 cm, halfway down the page and with its centre 8 cm from the left hand edge.
- Mark 20 points round the circumference of the circle.
- Mark a point, **A**, 2·5 cm from the left-hand edge level with the centre of the circle.

- With one of your points on the circumference as centre, draw a circle which passes through **A**.

- Repeat Step 2 for each of your other points.
- Draw the envelope.

(b) Investigate what happens when **A** lies on the circumference of the starting circle.
This envelope is called a **cardioid.**
Which part of the body is shaped like a cardioid?

Ask your teacher what to do next.

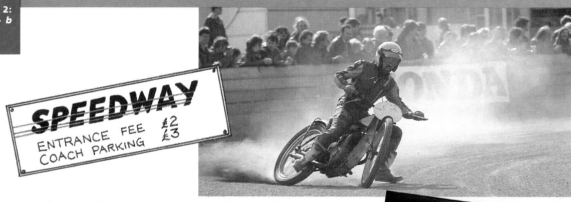

SPEEDWAY
ENTRANCE FEE £2
COACH PARKING £3

Jo is organising club outings and is costing different trips.

1 Thirty members want to visit the speedway.
Find **(a)** the total entrance fee
(b) the total cost of parking and entering.

2 Copy and complete.

Number in coach (n)	20	35	42	47	56
Entrance fees	40				
Total cost in pounds (c)	43				

This number machine shows the calculation.

number of people → ×2 → +3 → total cost

n × 2 + 3 = c

Formula   2n + 3 = c   or   c = 2n + 3

3 (a) Forty members want to visit the Waxworks. Find the total cost of parking and entering.
(b) Draw the number machine for this calculation.
(c) Write the formula.
(d) Use your formula to find the total cost for a coach party of 48 members.

GRAND U.K. TOUR
WAXWORKS
ENTRANCE FEE £2
COACH PARKING £4

4 Repeat question **3** for each of these events.

HORSE SHOW
Entrance Fee £4
Coach Parking £6

Entrance Fee £3
→ Air Show →
Coach Parking £5

ASSAULT COURSE
Entrance Fee £2
Coach Parking Free

STOCK CAR RACING

Entrance £1
Coach Parking £10

5 (a) Find a formula for the total cost of a coach party parking and visiting the Stock Car racing.
(b) Use your formula to find the total cost for a party of
• 23 • 32 • 46 • 53 members.

Supertrip pays Bee, the driver of their coach a basic weekly wage of £180 plus an extra £9 for each trip.
Bee checks her weekly wage using this formula

W = 180 + 9t where *W* is her wage in pounds
and *t* is the number of trips.

When she makes 5 trips in a week *t* = 5.

so   *W* = 180 + (9 × 5)
             W = 180 + 45
             W = 225

Her wage is £225.

6 Find Bee's wage in a week when she makes
 (a) 3 trips   **(b)** 4 trips   **(c)** 6 trips   **(d)** 7 trips.

7 George drives for Fastrip. He checks his wage using the formula
 W = 190 + 7*t*.
 Find his wage in a week when he makes
 (a) 2 trips   **(b)** 5 trips   **(c)** 7 trips   **(d)** 10 trips.

8 Jo gives 50p and each member gives 10p towards a collection for Bee.
 (a) Find the amount collected from twenty members and Jo.
 (b) Find a formula using *m* for the number of members
                     and *a* for the amount collected.
 (c) Use your formula to find the amount collected from Jo and
    • 25 members   • 30 members   • 42 members   • 53 members.

9 Some members want to visit the new ten-pin bowling alley.
 It costs £3 per person, but for groups of 30 or more there is a
 discount of £12.
 (a) Find the total cost for a group of 40 members.
 (b) Find a formula using *c* for the total cost in pounds
                     and *m* for the number of members.
 (c) Use your formula to find the cost for groups of
    • 34 members   • 37 members   • 49 members
    • 51 members   • 55 members   • 60 members.

Challenge

1 Angus would like the Club to visit the Highland Show.
The graph shows the total cost of parking plus entrance fees.

(a) Copy and complete the table.

Number of members (n)	2	5		10	
Total cost in £ (c)		26		35	

(b) The entrance fee is £3 each. What is the total entrance fee for a group of 10 members?

(c) How much is the parking fee?

(d) At what point does the graph cut the vertical axis?

(e) Copy and complete the formula
$c = \square\, n + \square$.

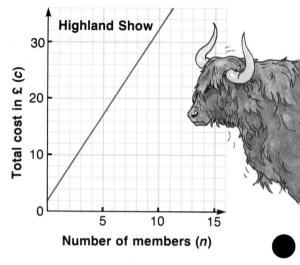

Highland Show

2 Morag wants to visit Atholl Castle. The graph shows the total cost of parking plus entrance fee.

(a) Copy and complete the table.

Number of members (n)	3	7		11	
Total cost in £ (c)			22		34

(b) Find the total cost for a group of seven members.

(c) The parking fee is £4. What is the entrance fee per member?

(d) At what point does the graph cut the vertical axis?

(e) Copy and complete the formula
$c = \square\, n + \square$.

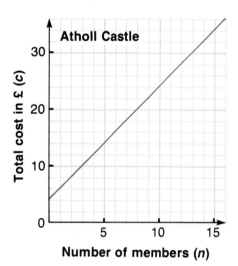

Atholl Castle

3 Fiona has always wanted to visit the Ice Palace. The graph shows the total cost of a visit.

(a) Copy and complete the table.

Number of members (n)	2	4		10	
Total cost in £ (c)		24		42	

(b) What is the • parking fee
• entrance fee?

(c) Copy and complete the formula
$c = \square\, n + \square$.

4 Explain how to find the parking fee for each of these graphs.

5 Do Workbook page 28.

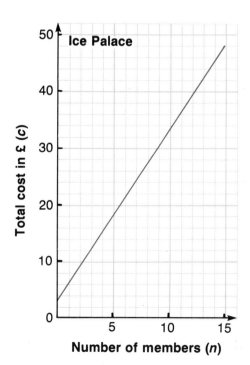

Ice Palace

1 Alan, in the box office, sends out tickets through the mailing
service. A theatre ticket weighs 3 grams and an envelope
weighs 5 grams. The formula for the weight of an envelope
and its contents is
$w = 3n + 5$ where w is the total weight in grams
             and n is the number of tickets.
Find the total weight of an envelope containing
(a) 2 tickets   **(b)** 5 tickets   **(c)** 7 tickets.

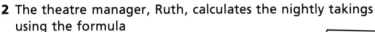

2 The theatre manager, Ruth, calculates the nightly takings
using the formula
    $t = 5b + 8s$ where t is the takings,
    b is the number of balcony tickets
and s is the number of stall tickets sold.
Calculate the takings if the theatre sells
(a) 120 balcony and 85 stall tickets
(b) 165 balcony and 105 stall tickets
(c) 138 balcony and 92 stall tickets.

Tickets for this
performance
Balcony £5
Stalls £8

3 The stagehand, Libby, frames posters for the foyer.
The formula for the perimeter of a poster is
$p = 2l + 2b$ where l is the length
             and b is the breadth.
How much wood does she need for posters with
(a) $l = 50$ cm, $b = 40$ cm   **(b)** $l = 60$ cm, $b = 45$ cm?

4 Brian, the theatre joiner, has to build frames for a new set design.
For each frame
• write a formula for the perimeter P
• find the value of the perimeter when $x = 4$ m and $y = 3$ m.

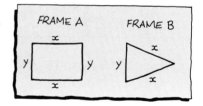

FRAME A   FRAME B

5 The set designer, Jolande, has to paint several large
triangular sheets of canvas.
Use the formula $A = \frac{1}{2}bh$ to find the area of each triangle,
when **(a)** $b = 4$ m, $h = 8$ m   **(b)** $b = 8$ m, $h = 3\cdot5$ m.

h

base (b)

Challenge

6 Jolande is planning to carpet the rectangular
stage, except for the trap door.
She has made this sketch.
(a) Write a formula for the area of
   • the whole stage (W)
   • the trap door (T)
   • the carpet (C).
(b) Use your formula to calculate C
   when $y = 12$ m, $x = 18$ m and $z = 2$ m.

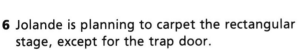

Plan of stage

y

Trap
door z

z

x

Ask your teacher what to do next.

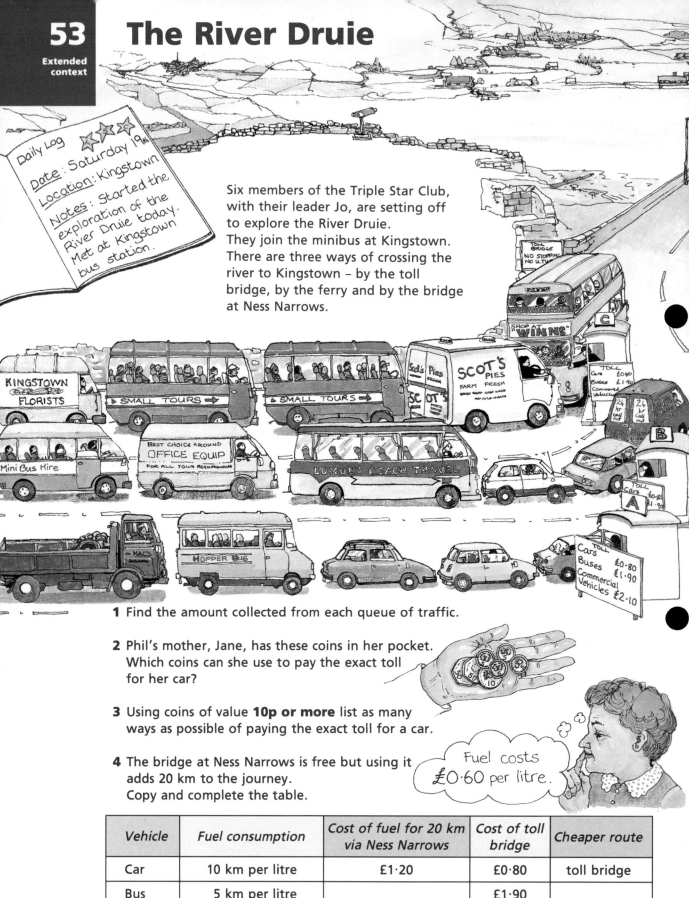

Six members of the Triple Star Club, with their leader Jo, are setting off to explore the River Druie.
They join the minibus at Kingstown.
There are three ways of crossing the river to Kingstown – by the toll bridge, by the ferry and by the bridge at Ness Narrows.

Daily Log ☆☆☆
Date: Saturday 19th
Location: Kingstown
Notes: Started the exploration of the River Druie today. Met at Kingstown bus station.

1 Find the amount collected from each queue of traffic.

2 Phil's mother, Jane, has these coins in her pocket. Which coins can she use to pay the exact toll for her car?

3 Using coins of value **10p or more** list as many ways as possible of paying the exact toll for a car.

4 The bridge at Ness Narrows is free but using it adds 20 km to the journey. Copy and complete the table.

Fuel costs £0·60 per litre.

Vehicle	Fuel consumption	Cost of fuel for 20 km via Ness Narrows	Cost of toll bridge	Cheaper route
Car	10 km per litre	£1·20	£0·80	toll bridge
Bus	5 km per litre		£1·90	
Van	8 km per litre			
Lorry	6 km per litre			

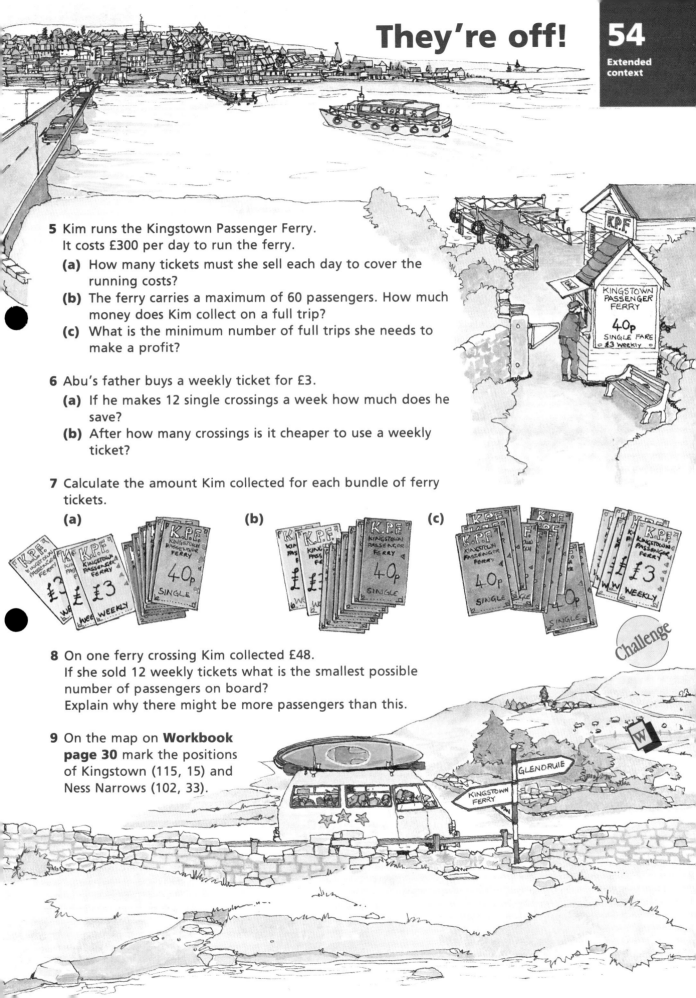

5 Kim runs the Kingstown Passenger Ferry.
It costs £300 per day to run the ferry.

  (a) How many tickets must she sell each day to cover the running costs?

  (b) The ferry carries a maximum of 60 passengers. How much money does Kim collect on a full trip?

  (c) What is the minimum number of full trips she needs to make a profit?

6 Abu's father buys a weekly ticket for £3.

  (a) If he makes 12 single crossings a week how much does he save?

  (b) After how many crossings is it cheaper to use a weekly ticket?

7 Calculate the amount Kim collected for each bundle of ferry tickets.

  (a)   **(b)**   **(c)**

8 On one ferry crossing Kim collected £48.
If she sold 12 weekly tickets what is the smallest possible number of passengers on board?
Explain why there might be more passengers than this.

Challenge

9 On the map on **Workbook page 30** mark the positions of Kingstown (115, 15) and Ness Narrows (102, 33).

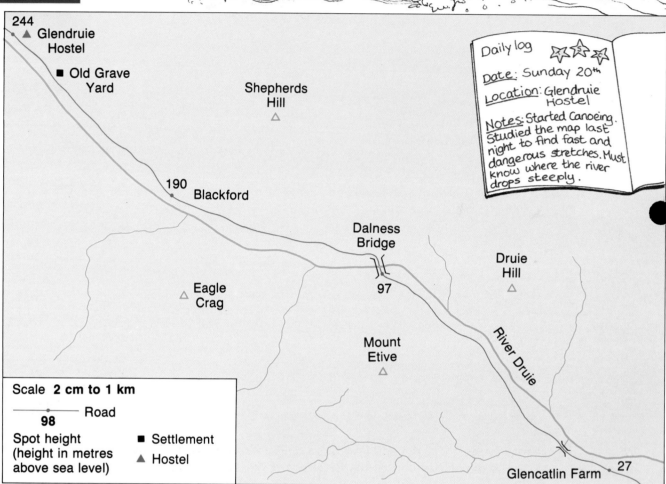

Scale 2 cm to 1 km

Road

98 Spot height (height in metres above sea level)

■ Settlement
▲ Hostel

Julie and Dai used the spot heights on the map to find the drop in the height of the river between two places. They also measured the distance between the two places.

1 (a) What is the height of the river at • Glendruie Hostel • Blackford?

(b) For the stretch of the river from the hostel to Blackford find
  • the drop in height • the distance in kilometres
  • the average drop in metres per kilometre (m/km).

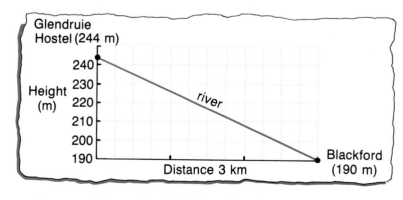

Julie drew this **longitudinal profile** to show how steeply the river drops.

2 (a) For the stretch from Blackford to Dalness Bridge find
  • the drop in height
  • the distance in km
  • the average drop in m/km.

(b) Using the same scales draw a longitudinal profile for this stretch of the river.

3 Repeat question **2** for the stretch from Dalness Bridge to Glencatlin Farm.

4 Which of the three stretches of river drops most steeply?

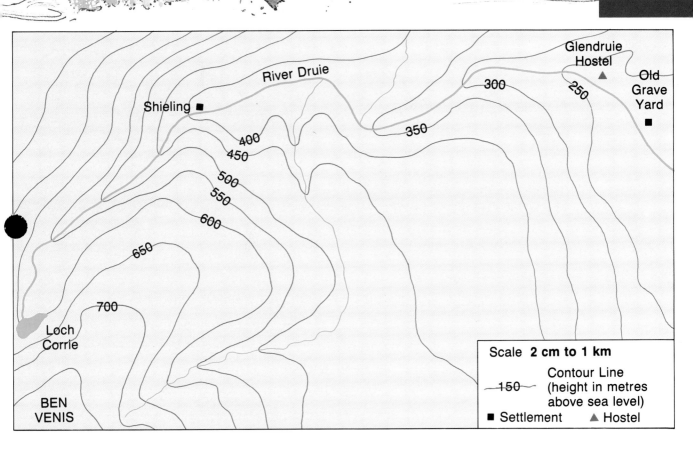

Where there are no spot heights Julie and Dai found heights from **contour lines**. These are the red lines on the map. They join points which are at the same height.

5 (a) What is the difference in height between neighbouring contour lines?
  (b) What is the height of   • Loch Corrie   • Shieling?
  (c) For the stretch from Loch Corrie to Shieling find the drop in height
  (d) The river twists a lot as it flows through this stretch.
     Find a way to measure the length on the map accurately.
  (e) Use the scale to calculate the actual distance the river flows.
  (f) Calculate the average drop.
  (g) Using the same scales as before draw a longitudinal profile.

6 Repeat question **5** for the stretch between Shieling and the Old Grave Yard
  (which is at a height of 225 m).

7 Hamish, the warden at Glendruie, showed
  Dai this profile of the maximum gradient
  likely to be safe for canoeing.
  (a) Use your profiles to decide which
     stretches of the river are not safe.
  (b) Suggest why the other stretches might
     not be safe.

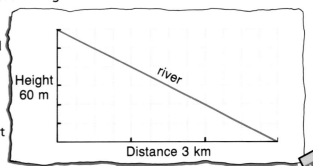

8 Mark these places on the map on **Workbook page 30.**
  • Loch Corrie (00,80) • Shieling (12, 92) • Glendruie Hostel (34,92) • Dalness Bridge (54,78)

Bridge over the Druie

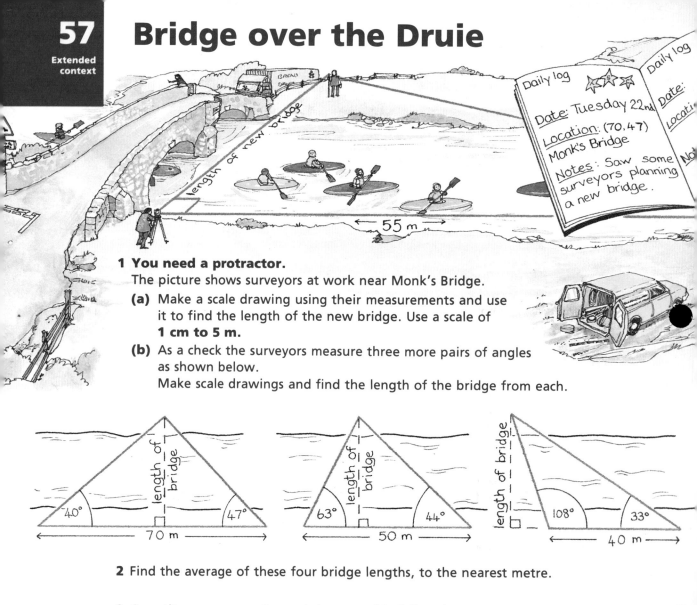

Daily log ✩✩✩
Date: Tuesday 22nd
Location: (70, 47)
Monk's Bridge
Notes: Saw some
surveyors planning
a new bridge.

Daily log
Date:
Locati
No

← 55 m →

1 You need a protractor.
The picture shows surveyors at work near Monk's Bridge.

(a) Make a scale drawing using their measurements and use
it to find the length of the new bridge. Use a scale of
1 cm to 5 m.

(b) As a check the surveyors measure three more pairs of angles
as shown below.
Make scale drawings and find the length of the bridge from each.

40° length of bridge 47°
← 70 m →

63° length of bridge 44°
← 50 m →

length of bridge 108° 33°
← 40 m →

2 Find the average of these four bridge lengths, to the nearest metre.

3 Quantity surveyors estimated the cost of building the new
bridge. Their table shows how the cost depends on its length.

Length (m)	5	10	15	20	25	30	35
Cost (£ thousands)	50	100	175	300	475	750	1125

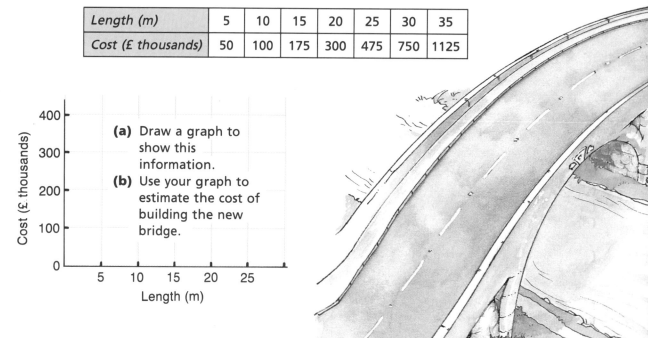

(a) Draw a graph to
show this
information.

(b) Use your graph to
estimate the cost of
building the new
bridge.

Cost (£ thousands) — axis: 0, 100, 200, 300, 400
Length (m) — axis: 5, 10, 15, 20, 25

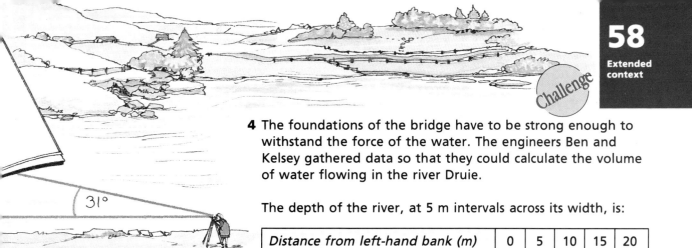

Challenge

4 The foundations of the bridge have to be strong enough to withstand the force of the water. The engineers Ben and Kelsey gathered data so that they could calculate the volume of water flowing in the river Druie.

The depth of the river, at 5 m intervals across its width, is:

Distance from left-hand bank (m)	0	5	10	15	20
Depth (m)	0	1·5	2·0	0·5	0

(a) Sue watched Kelsey drawing the **vertical profile** of the river. Copy and complete the profile.
(b) Calculate the area of each strip in the profile.
(c) Find the total area of the strips. This is called the **cross-sectional area A** of the river.

River Profile – Monk's Bridge

5m   5m   5m

5 Kelsey and Ben measured the speed of the water, on four occasions, by timing how long it took a float to travel a distance of 50 metres. Ben noted down the times.
1 min 42 s, 1 min 39 s, 1 min 38 s and 1 min 41 s.

(a) What was the average time in seconds?
(b) Calculate the **speed v** of the river in metres per second.
(c) The volume of water flowing down a river is called the **riverflow R**. It is calculated, in m³ per second, from the formula $R = Av$.
Calculate the riverflow at Monk's Bridge.

6 Ben and Kelsey calculated the average riverflow for each month over a year. Their results are shown in the bar graph.

Average riverflow – Monk's Bridge

Riverflow (m³ per second)

Jan Feb Mar Apr May Jun Jul Aug Sep Oct Nov Dec

(a) In which month was the riverflow least?
(b) What was the greatest value for the riverflow?
(c) What was the range of the monthly values?
(d) What was the average monthly value of the riverflow over the year?
(e) What value of riverflow should Kelsey and Ben design the new bridge to withstand? Explain your answer.

7 Mark the position of Monk's Bridge on the map on **Workbook page 30**.

Gone Fishing

1 Inside the River Druie Preservation Trust display centre the club members found lots of information about salmon. In which years were more salmon caught by rod than by net?

2 How many salmon were caught
   (a) by net, in 1986
   (b) by rod, in 1991
   (c) by net and rod together, in 1989?

3 Between which years did the **total number** of salmon caught
   (a) increase **(b)** decrease?

4 Was Ken correct when he said that there used to be a lot more salmon in the river?

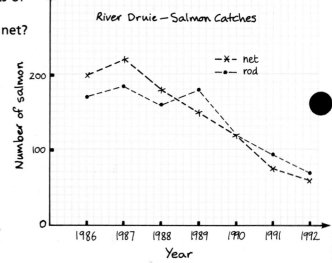

FISHY FACTS 1

During its life the main enemies of the salmon are birds, bigger fish, sharks and seals, but man is the greatest enemy of all.

RDPT

5 **(a)** A female salmon produces about 1250 eggs per kilogram of body weight. About how many eggs are produced by a female salmon weighing 4 kg?
   (b) The eggs develop into fish which stay in the river for two or three years. About 50 of these fish survive and swim to the sea. At this stage they are called **smolts.** What percentage of the **eggs** is this?
   (c) Only 10% of the **smolts** survive and return to the river as adult salmon. How many adults return?
   (d) What percentage of the original eggs develop into adults and return to the river?

6 Read Fishy Facts No 2.
How much higher or lower than your classroom
wall is the highest salmon leap?

FISHY FACTS 2

When returning
upstream the highest
leap a salmon has been
known to make in
Scotland is 3·65m.

RDPT

FISHY FACTS 3

The heaviest salmon
ever caught in Britain
weighed 29 kg. It was
caught in the River
Tay in 1922.

RDPT

7 (a) Last year in Scotland 25 000 anglers caught an average
of 2·4 salmon each. How many salmon did they catch?
(b) The total weight of these salmon was 240 000 kg.
What was the average weight of a salmon?
(c) About how many average weight salmon would be
needed to match **your** weight?

8 The salmon was sold at £8·75 per kg. What was the value of
(a) an average salmon
(b) all the salmon caught in the year?

9 How much would you cost if you were a salmon?

10 In Scotland last year the anglers spent about £48m on hotels,
travel, food and other expenses. If Ken's costs are typical,
how many days in the year on average did each angler spend
fishing?

My fishing
holiday
works out
at £60
a day!

Challenge

Challenge

11 You need ½ cm squared paper.
Draw this salmon enlarged by a scale factor of 4.

12 Mark the River Druie Preservation Trust's display
centre on the map on **Workbook page 30.**

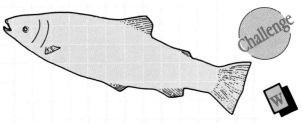

Warning: Pollution!

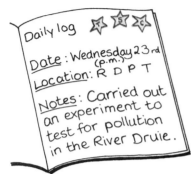

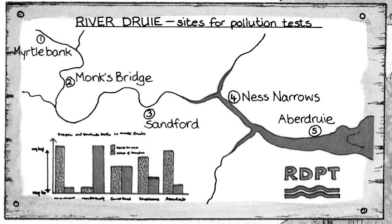

If the river becomes polluted the salmon can die. The
Preservation Trust tests for pollution at 5 sites. These are their
test results.

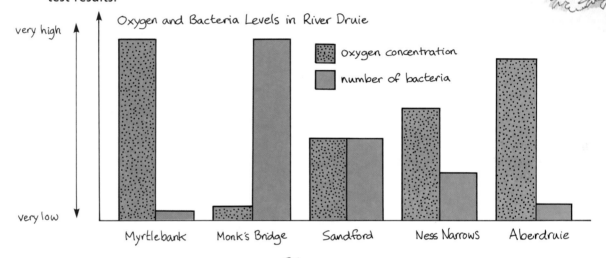

1 (a) There is **no** pollution at Myrtlebank.

Describe • the oxygen concentration there
• the number of bacteria there.

(b) At which of the other sites is there very little pollution?

2 (a) There is **very high** pollution at Monk's Bridge.

Describe • the oxygen concentration there
• the number of bacteria there.

(b) There is **high** pollution at Sandford.
Describe the pollution at Ness Narrows.

3 The main cause of the pollution in the River Druie is effluent
from sewage works. Bacteria feed on the organic waste. They
grow, multiply and absorb oxygen from the surrounding
water.

(a) Where does the effluent enter the River Druie?

(b) What happens to the oxygen concentration as the river
flows downstream past Sandford, Ness Narrows and
Aberdruie?

(c) Suggest a reason for this.

4 To obtain information about pollution, Fay and Dai used sweep nets to catch water animals. In these pictures each dimension of the water animals is **twice** its true size.

(a) Find the true body length of each in millimetres.

(b) List the animals which are shorter than the width of your thumb.

water louse

caddis fly larva

shrimp

skater

5 (a) Draw a bar graph to show the numbers of each type of water animal caught by Fay and Dai at Myrtlebank.

water louse – 1   shrimp – 2
caddis fly larva – 3   mayfly nymph – 12
stonefly nymph – 50

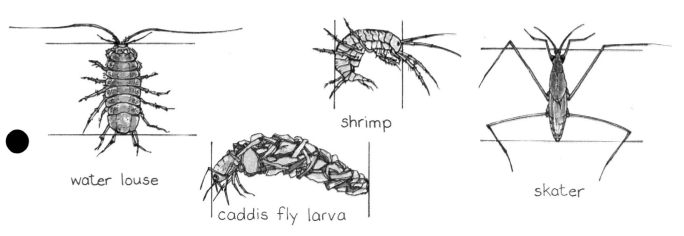

mayfly nymph   stonefly nymph

(b) What does the presence of large numbers of mayfly and stonefly nymphs show about water quality?

6 Mark these places on the map on **Workbook page 30.**

- Myrtlebank   (69,66)
- Aberdruie   (108,20)
- Sandford   (83,33)
- Sewage Works   (80,63)

Messing about on the river

Daily Log
Date: Friday 25th
Location: Donin
Notes:
Explored the mouth
of the river.

You need a protractor and Workbook page 29.

1 Look at the scale on the map.
   Find the actual distance represented by
   (a) 1 cm
   (b) 5 cm.

2 For each of the following pairs of places
   • measure the distance between them on the map
   • calculate the actual distance in metres.
   (a) Lighthouse and Donin
   (b) Pier and aquarium
   (c) Viewpoint and jetty.

3 Find the bearing of
   (a) the aquarium **from** the pier
   (b) the pier from the buoy
   (c) Craggan from the jetty
   (d) the Marine Centre from the lighthouse.

4 Sue's course from the pier to the Marine Centre
   is marked on the map. Complete the table in the
   workbook for her course.

	Bearing	Distance
First leg	322°	2000 m
Second leg	062°	3500 m
Third leg	356°	1000 m

5 Abu canoed from Sandylea to Portman
   and made a record of his course.
   Plot his course on the map.

6 The harbour master is planning to place a buoy on bearing
   260° from the lighthouse and 020° from the viewpoint.
   (a) Mark the position of the buoy.
   (b) How far is it from   • the lighthouse
                              • the viewpoint?

7 The group has arranged to meet the minibus at Donin for the
   journey home. Fay is at the lighthouse. What is the distance
   and bearing of her course to Donin?

8 Mark the positions of Donin (118,14) and Sandylea (123,6) on
   the map on **Workbook page 30.**

9 The Triple Star Club members write a report of their
   exploration. Imagine you are one of them and write about
   your trip down the River Druie.

Ask your teacher what to do next.

Tara wants to travel from King's Cross to Waterloo on the Underground. The fare is £1·50. The ticket machine only accepts £1 and 50p coins. There are three possible ways that Tara can insert the coins:

50p, £1   or   £1, 50p   or   50p, 50p, 50p

1 List all the possible ways of inserting coins for Underground fares of

(a) £1   (b) £2   (c) £2·50

2 Copy and complete this table to show the number of different ways of inserting coins.

Tube fare	50p	£1	£1·50	£2	£2·50
Number of ways			3		

3 Look carefully at the numbers in the bottom row of the table.
(a) Add   • the 1st and 2nd   • the 2nd and 3rd   • the 3rd and 4th numbers.
(b) Write about what you find.
(c) How many different ways could you pay for a £3 fare?   Check by listing.

To make a sequence like 1, 2, 3, 5, 8 . .
• Start with any two numbers.
• Add these to make the third number.
• Add the second and third to make the fourth and so on.

1, 2
1, 2, 3
1, 2, 3, 5
1, 2, 3, 5, 8

4 Write the next six numbers in sequences like this which start:
(a) 4, 5, 9, 14 . .   (b) 1, 3, 4 . .   (c) 5, 7 . .   (d) 25, 30 . .

The sequence 1, 1, 2, 3 . . is called the **Fibonacci sequence**.

5 Write the first ten numbers in the Fibonacci sequence.

6 Add   • the 1st, 2nd and 3rd numbers of the Fibonacci sequence.
        • the 2nd, 3rd and 4th numbers of the Fibonacci sequence.
Continue to add three consecutive numbers of the sequence.
Write about what you find.

7 Investigate this pattern using any three consecutive numbers from the Fibonacci sequence. Write about what you find.

$2 \times 5 = \square$ $3 \times 8 = \square$ $5 \times 13 = \square$ $8 \times 21 = \square$
$3^2 = \square$ $5^2 = \square$ $8^2 = \square$ $13^2 = \square$

Ask your teacher what to do next.

Agent Dublo Sven is on a mission behind enemy lines. He uses the FRA-PER-DEC code to send signals to base.

Leading Wren Pennyfarthing decodes the signal by
- converting between fractions, decimals and percentages
- using the decoding table to find the letter

CODE MANUAL FRA	CODE MANUAL PER	CODE MANUAL DEC

fraction → percentage
fraction → decimal

percentage → fraction
percentage → decimal

decimal → fraction
decimal → percentage

Examples

$\frac{29}{100}$ → 29%
$\frac{29}{100}$ → 0·29

$\frac{2}{5}$ → $\frac{4}{10}$ → $\frac{40}{100}$ → 40%
$5\overline{)2·0}$ → 0·4

Examples

53% → $\frac{53}{100}$
53% → 0·53

30% → $\frac{30}{100}$ → $\frac{3}{10}$
30% → 0·30 → 0·3

Examples

0·37 → $\frac{37}{100}$
0·37 → 37%

0·7 → $\frac{7}{10}$ → $\frac{70}{100}$
0·7 → 70%

page 9 · page 10 · page 11

1 Use the table to help decode the messages on the pad.

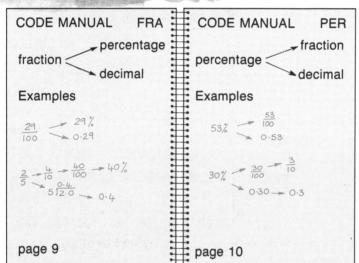

MESSAGE PAD

(a) 50%, 0·2, 0·5, $\frac{9}{10}$ $\frac{1}{3}$ // 30%, 40%, $\frac{3}{4}$, $\frac{2}{10}$ //
75%, $\frac{6}{10}$, 60%, $\frac{3}{4}$, $\frac{7}{10}$, 80% // 75%, $\frac{3}{5}$ //
90%, $\frac{1}{4}$, 1%, $\frac{1}{5}$, 0·25, $\frac{99}{100}$, 53%, $\frac{60}{100}$

(b) 0·15, 50%, 0·2, 1% // $\frac{90}{100}$, 100%, 35%, 0·5 //
$\frac{3}{4}$, $\frac{9}{10}$, 90%, 100%

(c) 23%, $\frac{1}{4}$, 0·4, $\frac{4}{10}$ // 77%, 40%, 100%, 0·23 //
$\frac{77}{100}$, $\frac{35}{100}$, 25%, 0·01, 0·99, 50% //
75%, $\frac{3}{5}$ // 0·19, $\frac{1}{4}$, 150%, 0·5

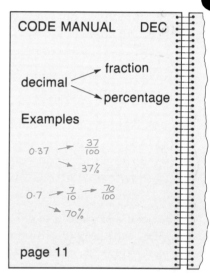

DECODING TABLE				
A	B	C	D	E
0·75	0·77	70%	$\frac{1}{100}$	$1\frac{1}{2}$
F	G	H	I	J
19%	99%	0·53	25%	11%
K	L	M	N	O
0·8	$\frac{2}{5}$	0·9	20%	1
P	Q	R	S	T
0·3	0·12	0·35	15%	0·6
U	V	W	X	Y
$\frac{1}{10}$	$1\frac{1}{2}$	$\frac{23}{100}$	$\frac{67}{100}$	$33\frac{1}{3}$%

2 Make up some messages for your partner to decode.

Tony is writing a report on the Secret Services.

TOP SECRET

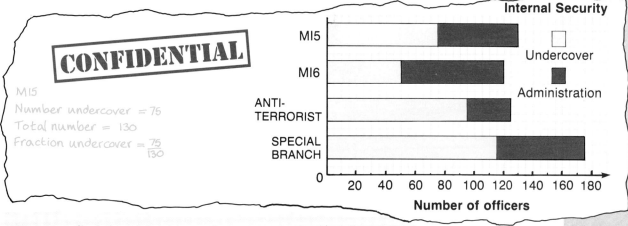

In 1950
Number of male agents = 160
Total number of agents = 200
The fraction who were male = $\frac{160}{200} = \frac{80}{100}$
The percentage who were male = 80%
The percentage who were female = 20%

Far East Section

1 For each year, find the percentage of agents who were
(a) male   **(b)** female.

CONFIDENTIAL

MI5
Number undercover = 75
Total number = 130
Fraction undercover = $\frac{75}{130}$

Internal Security

Tony can find the percentage of
MI5 undercover officers like this:

Enter **75** Press **÷ 1 3 0 =** to give **0.576923**
0·576923 is 0·58 **to the nearest hundredth.**

or Enter **75** Press **÷ 1 3 0 %** to give **57.692307**
57·692307 is 58 **to the nearest whole number.**

About **58%** of MI5 officers are undercover.

2 For each department find, to the nearest whole
number, the percentage of officers who are
(a) undercover   **(b)** administration.

PRIVATE

In 1940

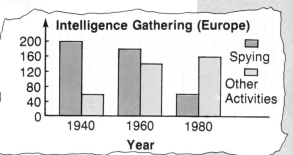

3 For each year find, to the nearest whole number,
the percentage of agents employed in
(a) spying   **(b)** other activities.

PART
THREE

Debbie is a dietician. She analyses the nutritional value of foods.

To find the weight of protein in Bertolini's Pasta Debbie calculates 12% of 140 g.

Enter `0.12` Press `×` `1` `4` `0` `=` to give `16.8`

or Enter `140` Press `×` `1` `2` `%` to give `16.8`

The weight of the protein is **16·8 g**.

1 For the packet of pasta calculate the weight of
   (a) carbohydrates   **(b)** fibre   **(c)** fat.

2 Find the weight of each nutrient listed in these items.
   (a)   **(b)**   **(c)**

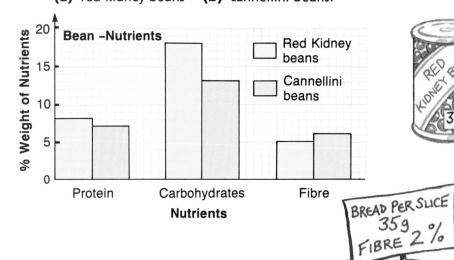

3 Debbie has drawn this graph to compare some of the nutritional content of two types of beans. Use the information to find the weight of each nutrient in the tin of
   (a) red kidney beans   **(b)** cannellini beans.

4 McGregors claim that a serving of their Muesli contains more fibre than 8 slices of bread.
Is this true? Explain.

Paul is increasing the prices of some items in stock.

1 For each item
  (a) calculate the increase in price
  (b) find the new price.

£326.50

INCREASE 4%

£450

INCREASE 24%

£62.75

INCREASE 5%

£278.80

INCREASE 17%

Laura takes over from Paul during his break. She has to increase the price of the midi-system.

The old price is 100%
$100\% + 7\% = 107\%$
$107\% = \dfrac{107}{100} = 1 \cdot 07$

INCREASE 36%

£89.50

INCREASE 7%

£570

She knows that the new price is **107% of £570.** Laura can find the new price like this:

Enter **1.07** Press **× 5 7 0 =** to give **609.9**
or Enter **570** Press **× 1 0 7 %** to give **609.9**

The new price of the midi system is **£609·90**.

2 Use Laura's method to check the new price of each of the items in question **1**.

3 Twenty people are presently employed at Ultra-tech. How many people will be employed during the sale period?

ULTRA-TECH

ULTRA-TECH ARE TO EMPLOY
15% MORE STAFF FOR THE
SALE PERIOD.

25 VIDEO TAPES

£43.75

4 Trevor has bought these video tapes from Ace Cash and Carry. He wants to make a profit of 52% on each tape. Find the selling price of each tape.

5 Shona works in the Ultra-tech office. She received this memo about a wage increase. Calculate, **to the nearest penny**, the new weekly wage for each of these employees.
  (a) Paul – assistant, weekly wage £256·48
  (b) Laura – supervisor, weekly wage £395·40
  (c) Trevor – manager, weekly wage £564·95.

ULTRA-TECH

Staff Memo - Pay Awards
The following pay rises will
take effect from this week.
Managers - 9%
Supervisors - 21%
Assistants - 12%

Athletica sell their discontinued lines to smaller sports shops. They give a discount on the Recommended Retail Price (RRP) of the items.

ATHLETICA SPORTS - DISCOUNT ORDER FOR			
STORE: Barney's		MONTH: August	
Item	RRP	Discount	Selling Pr
SWEATER	£48	15%	£40·80
SWIMSUIT	£26	13%	
TRAINERS	£42	5%	
TRACKSUIT	£66	23%	
T-SHIRT	£18·40	35%	
SHORTS	£11·25	16%	
SOCKS	£7·50	8%	

1 Barney has ordered these items from Athletica. For each item in the order

  (a) calculate the discount
  (b) find the selling price.

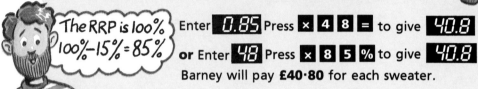

Barney knows a quick way to find the selling price.
He calculates the selling price of the sweater like this:

The RRP is 100%
100% – 15% = 85%

Enter **0.85** Press **×** **4** **8** **=** to give **40.8**

or Enter **48** Press **×** **8** **5** **%** to give **40.8**

Barney will pay **£40·80** for each sweater.

2 Use Barney's method to check the selling prices of the items in his order.

3 Find the sale price of each of these items.

£9·50

GOLF SHOES

£38·75

£348

ATHLETICA SPORTS
12% OFF GOLF STOCK for one day only all golf items in stock will be reduced by 12%

4 Athletica Sports employ 350 people. How many will they employ after the redundancies are made?

ATHLETICA SPORTS HIT BAD TIMES

The manager of Athletica Sports announced last night that 18% of the staff would be made redundant following the crisis in the

5 Find the figures for October when

  (a) the number of customers decreased by 13%
  (b) sales decreased by 9%.

Ask your teacher what to do next.

ATHLETICA SPORTS - MEMO
Monthly facts - September
Customers   32 600
Sales   £413 000

Robert sells flowers at Country Bunches.
The cost of one rose is *r* pence.
Carole's bunch of roses costs
$r + r + r + r + r = 5r$

Julie's bunch of roses costs
$r + r + r = 3r$

The **total** cost of these bunches of roses = $5r + 3r = 8r$
The **difference** in cost of these bunches = $5r - 3r = 2r$

1 Find each total.
 (a) $4r + 4r$ **(b)** $3r + 5r$ **(c)** $r + r$ **(d)** $21r + 15r$
 (e) $30r + 15r$ **(f)** $3r + r + 7r$ **(g)** $8r + 5r + 4r + 6r$

2 Find each difference.
 (a) $5r - 2r$ **(b)** $8r - 3r$ **(c)** $10r - 6r$
 (d) $12r - r$ **(e)** $6r - 5r$

3 Robert sells daffodils at *d* pence each. Find
 (a) $3d + 4d$ **(b)** $10d + 6d$ **(c)** $4d + 7d + 12d$
 (d) $7d - 2d$ **(e)** $9d - d$ **(f)** $16d - 10d$

4 Carnations cost *c* pence each. Find
 (a) $4c + 3c$ **(b)** $2c + c$ **(c)** $6c - 3c$
 (d) $4c + 5c - 2c$ **(e)** $7c + 4c - 2c$ **(f)** $10c - 4c + 3c$

The total cost of these bunches is
  $5c + 5d + 2c + 4d$
= $5c + 2c + 5d + 4d$
= $7c + 9d$

The difference in cost of these bunches is
  $5c + 5d - 2c - 4d$
= $5c - 2c + 5d - 4d$
= $3c + d$

5*c* and 2*c* are like terms.
5*d* and 4*d* are like terms.
You can only add and
subtract like terms.

5 Find
 (a) $3c + 2c + 3d + 6d$ **(b)** $5c + 3d + 4d$ **(c)** $6d + 4d + 5d$
 (d) $2c + 3r + 6c$ **(e)** $9c + 8d + 7c + 4d$ **(f)** $8r + 6c + 9c + 5r$
 (g) $7r + 2d - r$ **(h)** $5c + 6d - 3c - d$ **(i)** $9d + 8c - 4c - 2d$
 (j) $9r + 6d - 3r - d$ **(k)** $5r + 7c - 2r + 5c$ **(l)** $7c + 4r + 2c - 3r$
 (m) $8r + 2c - 3r + 7c$ **(n)** $6r + 4d - 3d - 2r$ **(o)** $6r + 3c - 6r - c$

Ask your teacher what to do next.

Any flat shape with 4 straight sides is called a **quadrilateral.**

1 Write the names of the quadrilaterals you know.

You need the table on Workbook page 32.

2 Measure the angles of each shape.
Write the sizes in the table.

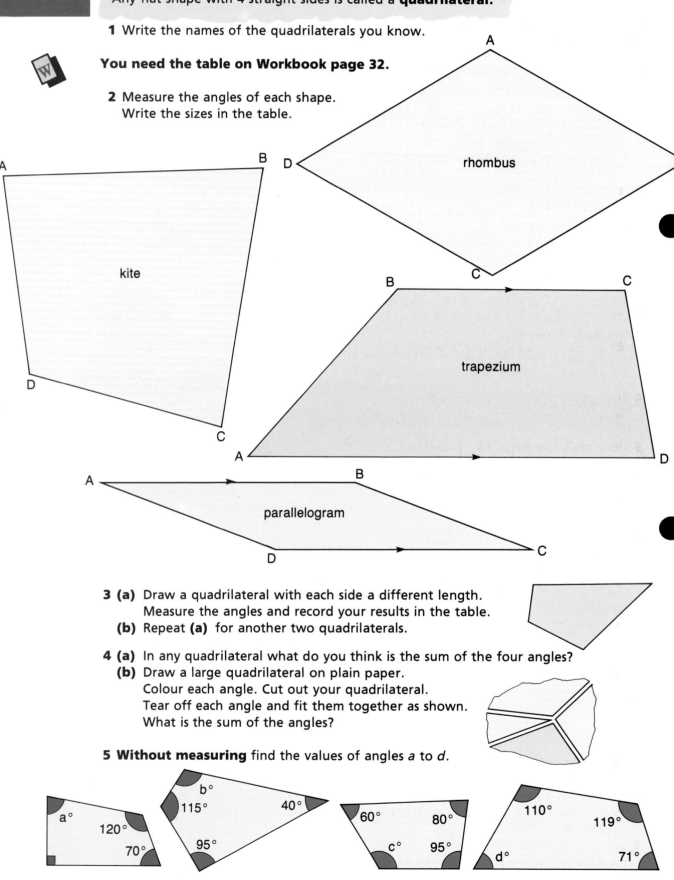

kite

rhombus

trapezium

parallelogram

3 (a) Draw a quadrilateral with each side a different length.
Measure the angles and record your results in the table.
(b) Repeat **(a)** for another two quadrilaterals.

4 (a) In any quadrilateral what do you think is the sum of the four angles?
(b) Draw a large quadrilateral on plain paper.
Colour each angle. Cut out your quadrilateral.
Tear off each angle and fit them together as shown.
What is the sum of the angles?

5 Without measuring find the values of angles *a* to *d*.

a°
120°
70°

b°
115°
40°
95°

60°
80°
c°
95°

110°
119°
d°
71°

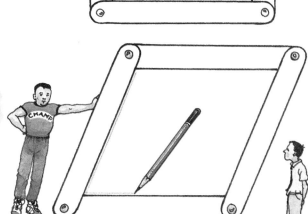

You need card, scissors, glue, paper fasteners and a protractor.

1 •   Cut off the bottom of **Workbook page 32** and stick it on card.
   •   Cut out each strip.
   •   Make holes for paper fasteners at the end of each strip.
   •   Keep your six strips to make shapes.

2 (a) Use paper fasteners to join 4 strips to make a square.
   (b) Push the corners of your square to make a shape like this.

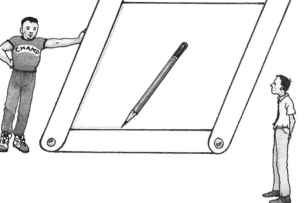

This shape is called a **rhombus.**

3 (a) Draw round the inside edges of your rhombus.
   (b) Measure the sides and angles in your drawing.

4 Push your rhombus a bit further.
Repeat question **3** for your new rhombus.

5 Do Workbook page 34, question 1.

6

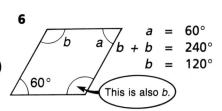

$a = 60°$ because the opposite angles of a rhombus are equal.
$b + b = 240°$ because the sum of the angles is 360°.
$b = 120°$

This is also b.

Sketch each rhombus. Without measuring fill in the sizes of **all** the sides and angles.

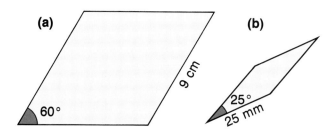

(a) 60°, 9 cm
(b) 25°, 25 mm

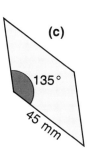

(c) 135°, 45 mm

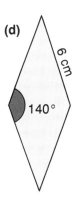

(d) 6 cm, 140°

7 Do Workbook page 34, question 2.

8 (a) Draw a rhombus starting with diagonals 10 cm and 6 cm.
   (b) Draw a rhombus starting with diagonals 8 cm and 5 cm.

9 Use tracing paper.
Investigate the line and rotational symmetry of a rhombus.

Investigation

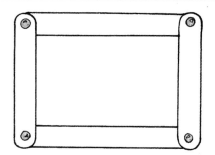

You need your six strips.

1 (a) Use paper fasteners to join 4 strips to make a rectangle.

(b) Push the corners of your rectangle to make a shape like this.

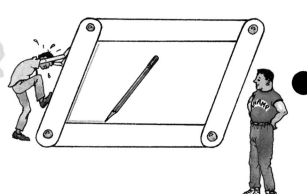

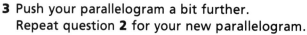
This shape is called a **parallelogram.**

2 (a) Draw round the inside edges of your parallelogram.
(b) Measure the sides and angles in your drawing.

3 Push your parallelogram a bit further.
Repeat question **2** for your new parallelogram.

4 Do Workbook page 33, question 1.

5 Sketch each parallelogram.
Without measuring, fill in the sizes of all the sides and angles.

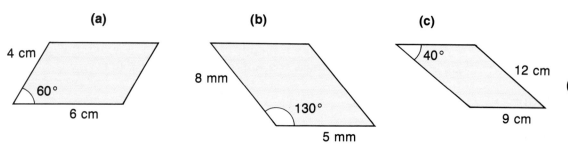

(a)

4 cm
60°
6 cm

(b)

8 mm
130°
5 mm

(c)

40°
12 cm
9 cm

6 Measure the diagonals of this parallelogram.
Measure AO, BO, CO, DO.
What do you notice?

7 Do Workbook page 33, question 2.

8 Draw three different parallelograms with diagonals 10 cm and 6 cm.

9 Investigate the symmetry of a parallelogram.

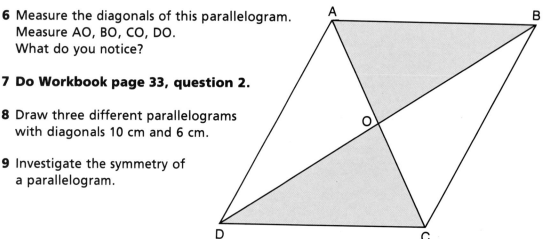

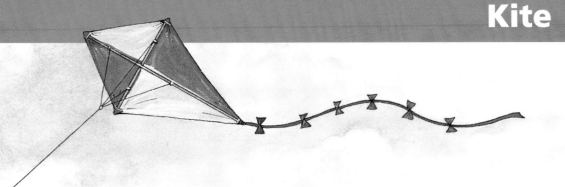

You need your six strips.

1 (a) Use paper fasteners to join 4 strips to make a shape like this.
 (b) Draw round the inside edges of your shape.
 (c) Measure the sides and angles in your drawing.

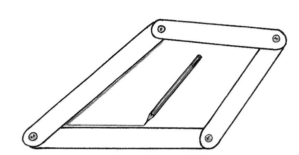

This shape is called a **kite**.

2 Push the corners of your **kite** to make another kite. Repeat question **1 (b)** and **(c)** for your new kite.

3 Push the corners further until you make this shape.

This shape is called a **V-kite**.

4 Do Workbook page 35.

5 Draw three different kites with diagonals 10 cm and 6 cm.

6 Sketch each kite.
 Without measuring, fill in the sizes of all the sides and marked angles.

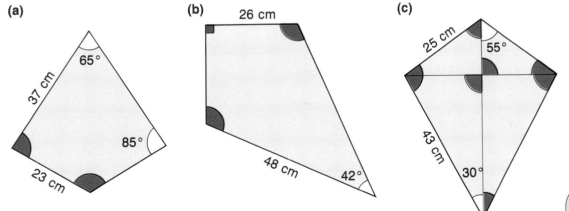

(a)

37 cm
65°
85°
23 cm

(b)
26 cm
48 cm
42°

(c)
25 cm
55°
43 cm
30°

7 Investigate the symmetry of a kite.

Investigation

8 Do Workbook page 32, question 2.

In reverse

1 Liz starts a number chain like this.

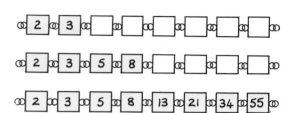

She continues the chain with a 5, then an 8.

Here is her complete chain.
Write Liz's rule for her number chain.

2 Do Workbook page 31.

3

Peter is solving Liz's number puzzles.

Find Liz's number for each of these puzzles.

(a)

I double my number and then add 4. The result is 16.

(b)

I multiply my number by 3 and then add 1. The result is 10.

(c)

I add 2 to my number and then double it. The result is 12.

(d)

I subtract 3 from my number and then multiply by 3. The result is 15.

4 The map shows Wemsby town centre.
Use the information to find each
person's **starting point**.

(a) Liz.   *Destination:* Bank
    Route: • second right
            • first left

(b) Peter.   *Destination:* Supermarket
    Route: • first left
            • second right

(c) Cathie. *Destination:* Cinema
    Route: • first left
            • second right
            • first left

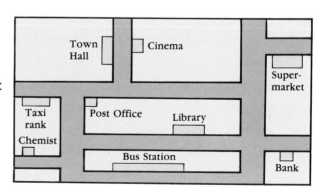

Ask your teacher what to do next.

1 Copy and extend this pattern of square numbers for the next three numbers.

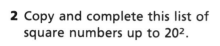

1   4   9

2 Copy and complete this list of square numbers up to 20^2.

$1^2 = 1 \times 1 = 1$
$2^2 = 2 \times 2 = 4$
$3^2 = 3 \times 3 = 9$
⋮
$20^2 =$

Remember

3^2 is read as three squared.

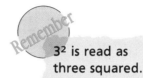

3 You can build up a pattern using square tiles.

shape 1

shape 2

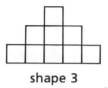

shape 3

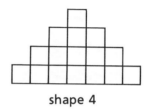

shape 4

(a) Draw the next 2 shapes in the pattern.
(b) Count the number of tiles in each shape and put your results in a table.
(c) How many tiles would be in shape   • 5   • 9   • 15?
(d) Without drawing, explain how to find the number of tiles when you know the number of the shape.

4 On isometric dot paper copy this pattern. Continue the pattern for the next two numbers.

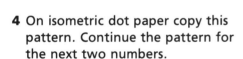

1   8   27

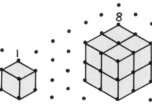
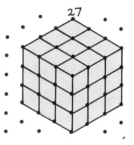

5 Copy and complete this list of cubic numbers up to 10^3.

$1^3 = 1 \times 1 \times 1 = 1$
$2^3 = 2 \times 2 \times 2 = 8$
$3^3 =$
⋮
$10^3 =$

Remember

2^3 is read as two cubed.

6 This pattern is built up using tiles.

1

shape 1

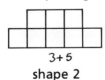

3 + 5

shape 2

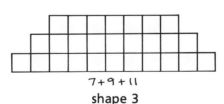

7 + 9 + 11

shape 3

(a) Draw the next shape in the pattern.
(b) Count the number of tiles in each shape and put your results in a table.
(c) How many tiles would be in shape   • 5   • 7   • 10?
(d) Without drawing, explain how to find the number of tiles when you know the number of the shape.

Rooting about

This square tile has an area of 64 cm².
What is the length of the side of the tile?

$$l^2 = 64$$
so $l = 8$

a number multiplied by itself is 64.
The number is 8.

We say 8 is the **square root of 64.**
The length of the tile is 8 cm.

1 Find the length of the side of each square tile.

(a)
36cm²

(b)
25cm²

(c)
49cm²

2 Write the **square root** of each of
these numbers.

(a) 4 **(b)** 9 **(c)** 16 **(d)** 81
(e) 100 **(f)** 121 **(g)** 144

3 In each list, find the square number
and write its square root.

(a) 7, 8, 9, 10, 11, 12
(b) 35, 36, 37, 38, 39, 40
(c) 46, 47, 48, 49, 50, 51
(d) 60, 61, 62, 63, 64, 65

You need a calculator.

4 (a) Enter **64.** Press ✓ and write your answer.
(b) Enter **9.** Press ✓ and write your answer.
(c) Enter **36.** Press ✓ and write your answer.
(d) What do you think ✓ means?

For the **square root of 64** we can write $\sqrt{64}$
$\sqrt{9}$ is read as the square root of 9. $\sqrt{9} = 3$
$\sqrt{36}$ is read as the square root of 36. $\sqrt{36} = 6$

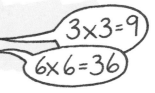
3×3=9
6×6=36

5 Find
(a) $\sqrt{4}$ **(b)** $\sqrt{25}$ **(c)** $\sqrt{1}$ **(d)** $\sqrt{81}$ **(e)** $\sqrt{64}$ **(f)** $\sqrt{100}$

6 Copy and complete.

Number	100	256	400	729	900	1024	1600	2401	2601
Square root									

Challenge

7 Write all the numbers less than 100 which have exact square roots.

8 Find all the numbers between 1200 and 1300 which have exact square roots.

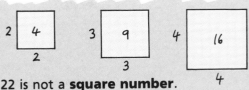
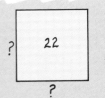

22 is not a **square number**.
$\sqrt{22}$ lies between 4 and 5.
From the graph $\sqrt{22} = 4\cdot7$.

1 Between which two whole numbers is the square root of

(a) 7   **(b)** 10   **(c)** 18   **(d)** 24
(e) 15   **(f)** 3   **(g)** 6   **(h)** 28?

2 Copy and complete
- $\sqrt{12}$ lies between ⬜ and ⬜.
- From the graph $\sqrt{12} = $ ⬜.

3 Repeat question **2** for

(a) $\sqrt{30}$ **(b)** $\sqrt{40}$ **(c)** $\sqrt{45}$
(d) $\sqrt{17}$ **(e)** $\sqrt{8}$ **(f)** $\sqrt{52}$
(g) $\sqrt{21}$ **(h)** $\sqrt{2}$ **(i)** $\sqrt{28}$

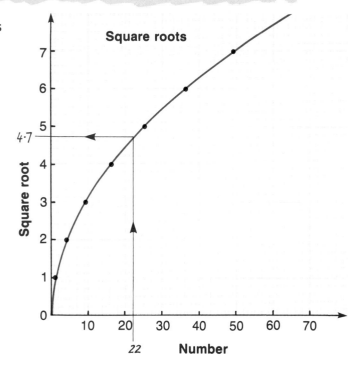

You need a calculator.

To find a better approximation for $\sqrt{22}$

Enter **22.** Press **√** to give **4.6904157**

$\sqrt{22} = 4\cdot69$ **correct to 2 decimal places.**

4 Find, correct to 2 decimal places

(a) $\sqrt{30}$ **(b)** $\sqrt{40}$ **(c)** $\sqrt{45}$ **(d)** $\sqrt{17}$ **(e)** $\sqrt{8}$
(f) $\sqrt{52}$ **(g)** $\sqrt{21}$ **(h)** $\sqrt{2}$ **(i)** $\sqrt{28}$
Compare these answers with your answers for question **3**.

5 ABCD is a square with area 80 m². APQR is a square with area 17 m². Find DR correct to 2 decimal places.

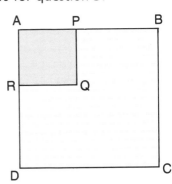

Challenge

1 Do Workbook page 36.

2 Copy and complete the number pattern.

$$100 = 10 \times 10 \qquad\qquad = 10^2$$
$$1\,000 = 10 \times 10 \times 10 \qquad\qquad = 10^3$$
$$10\,000 = 10 \times \qquad\qquad = 10^4$$
$$100\,000 = 10 \times \qquad\qquad =$$
$$1\,000\,000 = 10 \times \qquad\qquad =$$
$$10\,000\,000 = 10 \times 10 \times 10 \times 10 \times 10 \times 10 \times 10 =$$

> We read this as 10 to the power of 4.

3 Write the population of each town as a power of 10.

   (a) Brussels, 1 million
   (b) Mallaig, 1 thousand
   (c) Calcutta, 10 million
   (d) Ilfracombe, 10 thousand

4 Write each population as a power of 10.

   (a) Largs, 1 000
   (b) Cambridge, 100 000
   (c) Brisbane, 1 000 000
   (d) Buenos Aires, 10 000 000
   (e) Tewkesbury, 10 000
   (f) China, 1 000 000 000

5 Write each population as an ordinary number.

   (a) Birmingham, 10^6
   (b) Ullapool, 10^3
   (c) Blackburn, 10^5
   (d) Paris, 10^7
   (e) Corsham, 10^4
   (f) Raasay, 10^2

Liverpool has a population of 400 000.
This can be written as
$$4 \times 100\,000$$
$$= 4 \times 10^5$$

> This number must lie between 1 and 10.

The number is now written in **standard form**.

6 Write each population in standard form.

   (a) Athens, 3 000 000
   (b) Gloucester, 90 000
   (c) Iran, 40 million
   (d) Newcastle, 200 000
   (e) Margate, 50 thousand
   (f) India, 700 000 000

Toronto has a population of 3×10^6
$$= 3 \times 1\,000\,000$$
$$= 3\,000\,000$$

7 Write each population as an ordinary number.

   (a) Bristol, 4×10^5
   (b) West Yorkshire, 2×10^6
   (c) East Kilbride, 8×10^4
   (d) Oban, 8×10^3
   (e) Fort Augustus, 6×10^2
   (f) Argentina, 3×10^7

8 List these populations in order starting with the smallest.
   • Rome, 3 million   • Budapest, 2×10^6 • Lisbon, 800 000

Ask your teacher what to do next.

Image of Perth, Western Australia taken from a satellite. The buildings appear as blue/grey. The bright red area to the east of Perth is wooded upland.

Play each of these board games with a partner.

For each game one player needs three red counters and the other three blue counters.
Take it in turns to place a counter on the board.
Once all six counters have been placed take it in turns to move a counter along a line to an empty spot.

Game 1   Line Up

This board game comes from ancient China and is call Yih.

The winner is the first player to make a straight line with his/her three counters.

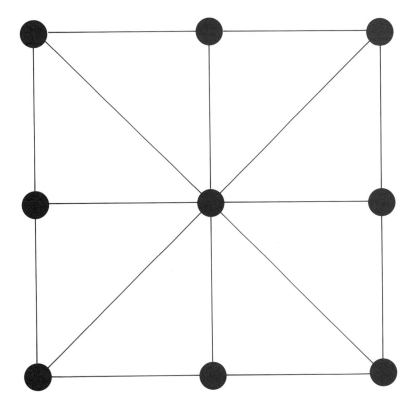

Game 2   Trapped

The winner is the first player to box in his/her partner so that he/she cannot move.

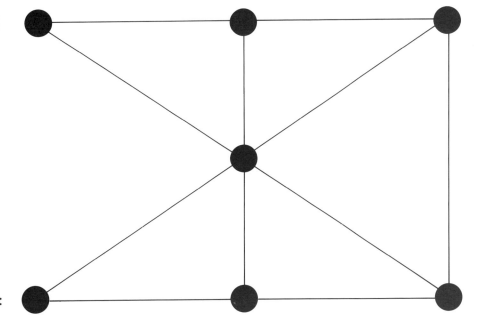

Ask your teacher what to do next.

The graph shows the petrol consumption for Bob's car.

The scale on the horizontal axis is **1 cm represents 5 litres.**

1 What is the scale on the vertical axis?

2 How far does the car travel on
   (a) 20 litres   **(b)** 15 litres
   (c) 7 litres   **(d)** 28 litres?

3 How much petrol is used for a trip of
   (a) 170 km   **(b)** 30 km
   (c) 85 km   **(d)** 115 km?

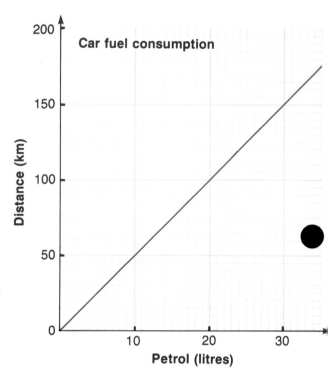

The graph shows the petrol consumption for Gill's motorcycle.

4 What is the scale on
   (a) the horizontal axis
   (b) the vertical axis?

5 How far does the bike travel on
   (a) 2 litres   **(b)** 5·6 litres
   (c) 3·2 litres   **(d)** 5 litres?

6 How much petrol is used for a trip of
   (a) 120 km   **(b)** 84 km
   (c) 36 km   **(d)** 156 km?

7 Which vehicle has the higher fuel consumption? Explain.

8 Do Workbook page 37.

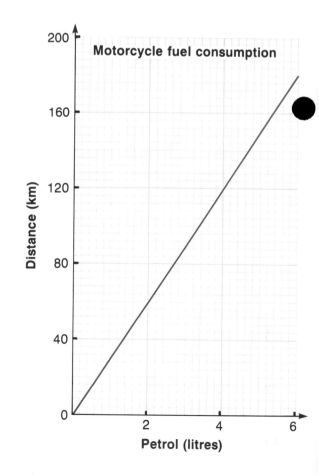

Petra boiled a beaker of water for a science experiment. She placed the beaker near an open window to cool and recorded the water temperature every 10 minutes. The graph shows her results.

1 What does 1 cm represent on
 (a) the horizontal axis
 (b) the vertical axis?

2 Estimate the water temperature after
 (a) 10 minutes **(b)** 25 minutes
 (c) 40, minutes **(d)** 55 minutes.

3 How long did the water take to reach a temperature of
 (a) 70°C **(b)** 52°C **(c)** 25°C **(d)** 16°C?

4 The water temperature did not fall below 16°C. Explain.

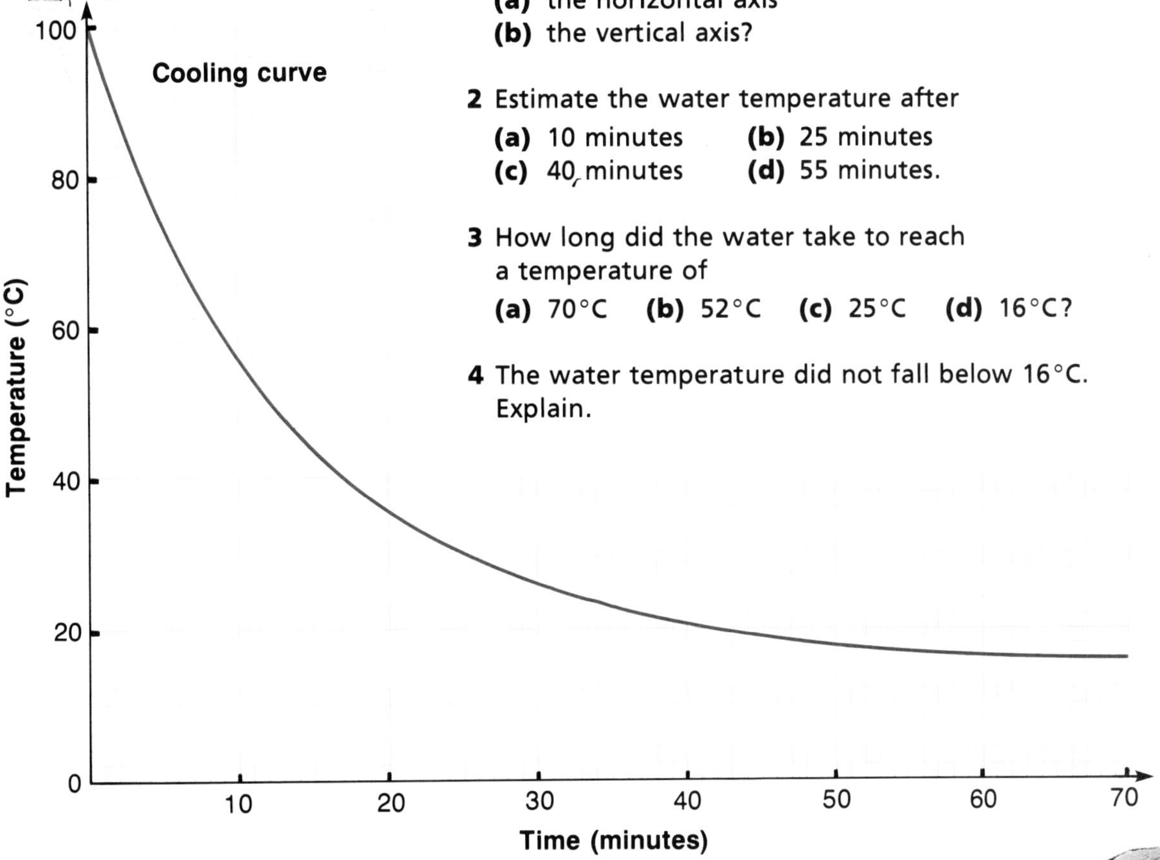

Cooling curve

Temperature (°C) — vertical axis: 0, 20, 40, 60, 80, 100
Time (minutes) — horizontal axis: 10, 20, 30, 40, 50, 60, 70

5 Shona carried out the same experiment at the other end of the lab. Here are her results.

Time (minutes)	0	10	20	30	40	50	60	70
Temperature (°C)	100	60	40	30	24	22	20	20

Use these scales to draw a graph of Shona's results.
Horizontal axis: 1 cm represents 5 minutes.
Vertical axis: 1 cm represents 5°C.

6 Use your graph to estimate the water temperature after
 (a) 4 minutes **(b)** 24 minutes.

7 How long did the water take to reach a temperature of
 (a) 52°C **(b)** 28°C **(c)** 20°C?

Ask your teacher what to do next.

1 These prisms are built from centimetre cubes and half cubes.

(a) Copy and complete this table.

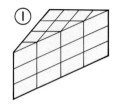
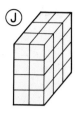

	Number of cubes in one layer	Number of layers	Volume in cubes
Prism Ⓘ	8	3	24
Prism Ⓙ			

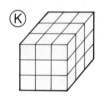
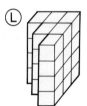
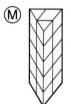
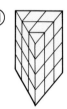

(b) Explain how you found the volumes of these prisms.

2 (a) These prisms are built from centimetre cubes and half cubes, and wrapped in coloured paper. Copy and complete this table.

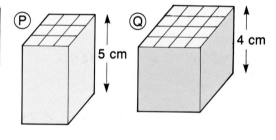

	Area of base A	Height h	Volume V
Prism Ⓟ	9 cm²	5 cm	
Prism Ⓠ			
Prism Ⓡ			

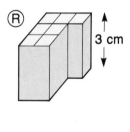
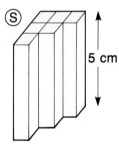
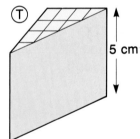
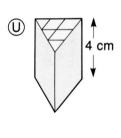

(b) Explain how you found the volumes of these prisms.

3 Copy and complete this formula for the volume of a prism.
Volume = Area of base × ☐

4 Calculate the volume of each of these prisms using the formula $V = Ah$.

(a)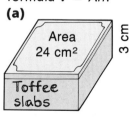

Area 24 cm²
Toffee slabs
3 cm

(b)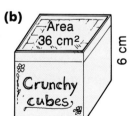

Area 36 cm²
Crunchy cubes
6 cm

(c)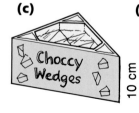

Choccy Wedges
10 cm
Area of base = 25 cm²

(d)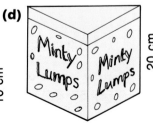

Minty Lumps
20 cm
Area of base = 60 cm²

Each of the tents in the Lomond Catalogue is a **triangular prism.**

You can find the volume V of air inside the 'FORCE 8' tent using:

V = Area of end × length

* First, find the area A of the end of the tent.

  $A = \frac{1}{2}bh$
  $\quad = \frac{1}{2} \times 4 \times 3$
  $\quad = 6\ m^2$

* Then find the volume.

  V = Area of end × length
  $\quad = 6 \times 5$
  $\quad = \textbf{30 m}^3$

The volume of air in the 'FORCE 8' tent is **30 cubic metres.**

1 Find the volume of air in each of the other tents.

2 Find the volume of each of these triangular prisms.

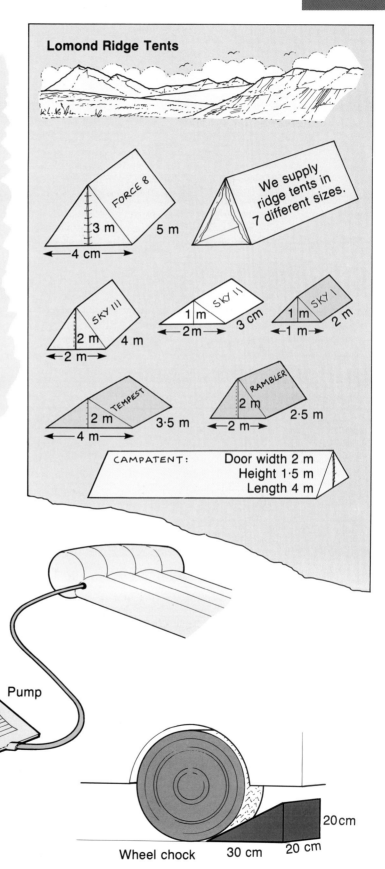

Lomond Ridge Tents

FORCE 8 — 3 m, 5 m, 4 cm

We supply ridge tents in 7 different sizes.

SKY III — 2 m, 4 m, 2 m

SKY II — 1 m, 2 m, 3 cm

SKY I — 1 m, 1 m, 2 m

TEMPEST — 2 m, 4 m, 3·5 m

RAMBLER — 2 m, 2 m, 2·5 m

CAMPATENT: Door width 2 m
Height 1·5 m
Length 4 m

cheese — 10 cm, 6 cm, 12 cm

Pump — 20 cm, 16 cm, 24 cm

Axe head — 4 cm, 10 cm, 8 cm

Wheel chock — 20 cm, 30 cm, 20 cm

Silver Arrow Carriers is a transport company. Rhaji calculates the carriage space or **capacity** of each size of van.

She calculates the capacity of the Atlas Van in two parts.

The capacity of the red part = 4 × 2 × 3 = 24 m³	The capacity of the yellow part = 1 × 2 × 1 = 2 m³

The total capacity = 24 + 2 = **26 m³**

2 m
1 m
3 m
1 m
2 m
4 m

1 Find the capacity, in m³, of each van.

The Portus Van

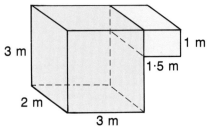

3 m
1 m
1·5 m
2 m
3 m

The Hercules Van

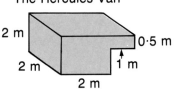

2 m
0·5 m
2 m
1 m
2 m

The Starlifter Van

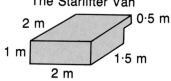

2 m
0·5 m
1 m
1·5 m
2 m

2 The ramp to the yard is made of concrete.
Find the volume of concrete in cubic metres.

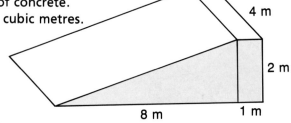

4 m
2 m
8 m
1 m

3 Find the volume of the Grit Bin.

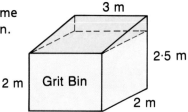

3 m
2·5 m
2 m Grit Bin
2 m

4 Lloyd, one of the drivers, delivers these wooden packing cases.
Find the volume of each.

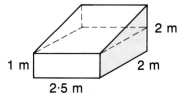

2 m
1 m
2 m
2·5 m

1 m
3 m
2 m
1·8 m

Ask your teacher what to do next.

A cylinder, a triangular prism, and a square pyramid are on the table.

This is Marek's view or **elevation**.

He sees the square pyramid on his right and the cylinder on his left.

1 Who can see each of these elevations?

(a)

(b)

(c)

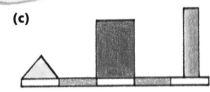

2 The cylinder is moved to this position.

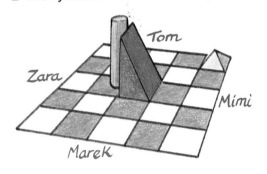

This is Mimi's elevation.

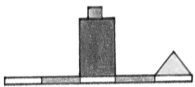

Draw and label the elevations seen by each of the others.

3 Draw and label each person's elevation for these positions.

(a)

(b)

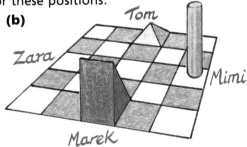

4 The shapes are moved again to give these elevations.

Marek's elevation

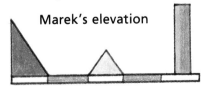

Mimi's elevation

Draw the elevations for Zara and Tom.

Ask your teacher what to do next.

The ladders outside the Jupiter rocket are numbered to help Jill the technician locate parts of the rocket which need maintenance.

> Jill starts at $^-8$ and goes up 5 rungs to reach $^-3$.
> We can write this as an addition.
> $^-8 + 5 = {}^-3$

1 Jill starts at $^-2$ and goes up 7 rungs.
  (a) What rung does she reach?
  (b) Copy and complete $^-2 + 7 = \square$.

2 Write additions to find which rung Jill reaches if she starts at
  (a) $^-7$ and goes up 3 rungs
  (b) $^-3$ and goes up 12 rungs
  (c) $^-4$ and goes up 4 rungs
  (d) $^-10$ and goes up 6 rungs.

3 Find   **(a)** $^-8 + 2$ **(b)** $^-1 + 7$ **(c)** $^-7 + 3$
         (d) $^-6 + 6$ **(e)** $^-4 + 8$ **(f)** $^-9 + 8$

4 Jill climbs from $^-2$ up to 4.
  (a) How far up does she climb?
  (b) Copy and complete: $^-2 + \square = 4$

5 Write additions to show the number of rungs she climbs if she goes up from
  (a) 1 to 8   **(b)** $^-9$ to 0   **(c)** $^-4$ to $^-1$
  (d) $^-9$ to 2   **(e)** $^-7$ to 4   **(f)** $^-3$ to 10

> Jill starts at 2 and goes down 5 rungs to reach $^-3$.
> We can write this as a subtraction.
> $2 - 5 = {}^-3$

6 Jill starts at $^-3$ and goes down 7 rungs.
  (a) What rung does she reach?   **(b)** Copy and complete: $^-3 - 7 = \square$.

7 Write subtractions to find which rung Jill reaches if she starts at
  (a) 6 and goes down 9 rungs   **(b)** $^-2$ and goes down 5 rungs
  (c) 9 and goes down 16 rungs   **(d)** $^-6$ and goes down 3 rungs.

8 Find   **(a)** $3 - 7$ **(b)** $5 - 8$ **(c)** $2 - 9$ **(d)** $^-1 - 5$ **(e)** $3 - 8$
         (f) $5 - 5$ **(g)** $8 - 5$ **(h)** $^-7 - 1$ **(i)** $^-2 - 7$ **(j)** $^-1 - 9$

9 Jill climbs from $^-3$ down to $^-8$.
  (a) How far down does she climb?   **(b)** Copy and complete: $^-3 - \square = {}^-8$.

10 Write subtractions to show the number of rungs climbed if she goes down
  from   **(a)** 7 to 0   **(b)** 5 to $^-5$ **(c)** 9 to $^-2$ **(d)** 6 to $^-10$

11 Find   **(a)** $^-3 + 8$ **(b)** $2 - 9$ **(c)** $^-8 + 2$ **(d)** $0 - 8$ **(e)** $^-10 + 3$
          (f) $5 - 17$ **(g)** $^-2 - 11$ **(h)** $^-18 + 7$ **(i)** $6 - 21$ **(j)** $^-30 - 9$

$$^-10 \quad ^-9 \quad ^-8 \quad ^-7 \quad ^-6 \quad ^-5 \quad ^-4 \quad ^-3 \quad ^-2 \quad ^-1 \quad 0 \quad 1 \quad 2 \quad 3 \quad 4 \quad 5 \quad 6 \quad 7 \quad 8 \quad 9 \quad 10$$

At the Disco Stars Dancing Competition the judges give points for **technique** and for **artistic expression.**

1 Zoe scores $^-6$ and 8.
 (a) What is her total score?
 (b) Copy and complete: $^-6 + 8 = \square$.

2 Freda scores 8 and $^-6$.
 She has the same total score as Zoe.
 Copy and complete: $8 + \,^-6 = \square$.

3 Copy and complete the additions to find the total score for each of these contestants:

Felix $6 + \,^-2$ Kay $5 + \,^-7$

     $= \,^-2 + \ 6$ 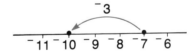 $= \,^-7 + \ 5$
    $= \square$ $= \square$

4 Find **(a)** $6 - 2$ **(b)** $5 - 7$

Look at your answers to question **3** and question **4**.
$6 + \,^-2 = 4$ adding $^-2$ is the same as subtracting 2
$6 - \ 2 = 4$

$5 + \,^-7 = \,^-2$ adding $^-7$ is the same as subtracting 7
$5 - \ 7 = \,^-2$

Adding a negative number is the same as subtracting the positive number.

5 Copy and complete the judge's calculation for Pip's total score.

Pip $^-7 + \,^-3$
    $= \,^-7 - \ 3$
    $= \square$

$^-3$
$^-11 \quad ^-10 \quad ^-9 \quad ^-8 \quad ^-7 \quad ^-6$

6 Copy and complete to find these total scores:

Sue $7 + \,^-2$ Pat $4 + \,^-5$ Kim $^-8 + \,^-1$

    $= 7 - \square$ $= 4 \square \square$ $= \square \square \square$
    $= \square$ $= \square$ $= \square$

7 Calculate the total score for each of these contestants.

Ria   Flo   Wes   Abu

8 Find **(a)** $^-7 + \,^-3$ **(b)** $2 + \,^-9$ **(c)** $8 + \,^-14$ **(d)** $7 + \,^-12$ **(e)** $5 + \,^-13$
     (f) $3 + \,^-9$ **(g)** $^-1 + \,^-8$ **(h)** $^-4 + \,^-20$ **(i)** $9 + \,^-35$ **(j)** $^-15 + \,^-40$

9 In the team event points are given for **teamwork** as well as **technique** and **artistic impression.**
 Calculate the total score for each of these teams.

Rappers   Boppers   Jivers

Jammers   Swingers   Breakers

Brenda completed a round in 73 strokes:
Her score is 3 **over** par or $^+3$.

Mac completed a round in 68 strokes:
His score is 2 **under** par or $^-2$.

Bryan's score after round 1 and round 2 was
$^-1 + {}^-1 = {}^-2$ so $2 \times {}^-1 = {}^-2$ **or** $^-1 \times 2 = {}^-2$

Bryan's score after 3 rounds was
$^-1 + {}^-1 + {}^-1 = {}^-3$ so $3 \times {}^-1 = {}^-3$ **or** $^-1 \times 3 = {}^-3$

Caiystane Golf Club				
18 holes				par 70
Round	*1*	*2*	*3*	*4*
Bryan	$^-1$	$^-1$	$^-1$	$^-1$
Val	$^-2$	$^-2$	$^-2$	$^-2$
Kim	$^-3$	$^-3$	$^-3$	$^-3$

1 Copy and complete for Bryan's score after 4 rounds.
$^-1 + {}^-1 + {}^-1 + {}^-1 = \square$ so $4 \times {}^-1 = \square$ or $^-1 \times 4 = \square$

2 Copy and complete for Val's score after
   (a) 2 rounds: $^-2 + {}^-2$ $= \square$ so $2 \times {}^-2 = \square$ or $^-2 \times 2 = \square$
   (b) 3 rounds: $^-2 + {}^-2 + {}^-2$ $= \square$ so $3 \times {}^-2 = \square$ or $^-2 \times 3 = \square$
   (c) 4 rounds: $^-2 + {}^-2 + {}^-2 + {}^-2$ $= \square$ so $4 \times {}^-2 = \square$ or $^-2 \times 4 = \square$

3 Copy and complete for Kim's score after
   (a) 2 rounds: $^-3 + {}^-3$ $= \square$ so $2 \times {}^-3 = \square$ or $^-3 \times 2 = \square$
   (b) 3 rounds: $^-3 + {}^-3 + {}^-3$ $= \square$ so $3 \times {}^-3 = \square$ or $^-3 \times 3 = \square$
   (c) 4 rounds: $^-3 + {}^-3 + {}^-3 + {}^-3$ $= \square$ so $4 \times {}^-3 = \square$ or $^-3 \times 4 = \square$

A **positive number** multiplied by a **negative number** gives a **negative number**.
A **negative number** multiplied by a **positive number** gives a **negative number**.

4 Find
   (a) $5 \times {}^-2$ **(b)** $^-2 \times 6$ **(c)** $7 \times {}^-2$ **(d)** $^-3 \times 5$
   (e) $6 \times {}^-3$ **(f)** $^-3 \times 7$ **(g)** $4 \times {}^-5$ **(h)** $^-6 \times 3$
   (i) $^-8 \times 2$ **(j)** $4 \times {}^-3$ **(k)** $5 \times {}^-5$ **(l)** $^-8 \times 3$
   (m) $^-7 \times 3$ **(n)** $^-4 \times 8$ **(o)** $^-2 \times 9$ **(p)** $^-6 \times 7$

Paula scored a total of $^-8$
over 4 rounds.
What was her **average** score?

	Total score	Number of rounds
Paula	$^-8$	4
Jane	$^-10$	2
Bert	$^-9$	3
Amos	$^-4$	4
Ella	$^-12$	4

$4 \times ? = {}^-8$
$4 \times {}^-2 = {}^-8$ $^-8 \div 4 = {}^-2$

Her average score was $^-2$ strokes.

5 Copy and complete to find the other average scores.
   (a) Jane: $^-10 \div 2 = \square$ **(b)** Bert: $^-9 \div 3 = \square$
   (c) Amos: $^-4 \div 4 = \square$ **(d)** Ella: $^-12 \div 4 = \square$

A **negative number** divided by a **positive number** gives a **negative number**.

6 Find
   (a) $^-6 \div 3$ **(b)** $^-8 \div 2$ **(c)** $^-6 \div 1$ **(d)** $^-15 \div 3$
   (e) $^-20 \div 5$ **(f)** $^-21 \div 7$ **(g)** $^-40 \div 5$ **(h)** $^-36 \div 6$
   (i) $^-45 \div 9$ **(j)** $^-18 \div 2$ **(k)** $^-30 \div 10$ **(l)** $^-26 \div 2$
   (m) $^-56 \div 4$ **(n)** $^-60 \div 3$ **(o)** $^-55 \div 5$ **(p)** $^-64 \div 8$

Tanya uses a coordinate diagram to help her design symmetrical patterns.

1 On this pattern, point A has coordinates (⁻3, 4).
Write the coordinates of points B to L.

2 For this pattern write the coordinates of the points which lie
  (a) on the y-axis
  (b) to the right of the y-axis
  (c) to the left of the y-axis
  (d) on the x-axis
  (e) above the x-axis
  (f) below the x-axis.

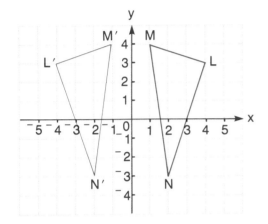

3 For this pattern, Tanya reflects the point L (4, 3) in the **y-axis.**
Its **image** is L′ (⁻4, 3).
Write the coordinates of
  (a) M and its image M′
  (b) N and its image N′.

4 Do Workbook page 38, questions 1 and 2.

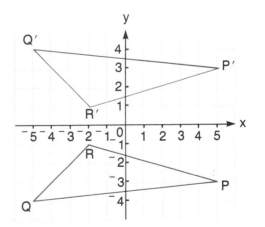

5 Use a separate coordinate diagram for each pattern. For each
  • plot the points and complete the shape.
  • reflect the shape in the **y-axis.**
  • complete the pattern.
  (a) D (1, 2), E (3, 4), F (4, ⁻3)
  (b) J (⁻2, 5), K (0, ⁻2), L (⁻3, ⁻3)

6 For this pattern, Tanya reflects the point P (5, ⁻3) in the **x-axis.**
Its **image** is P′ (5, 3).
Write the coordinates of
  (a) Q and its image Q′
  (b) R and its image R′.

7 Do Workbook page 38, questions 3 and 4.

8 Use a separate coordinate diagram for each pattern. For each
  • plot the points and complete the shape.
  • reflect the shape in the **x-axis**.
  • complete the pattern.
  (a) U (5, 1), V (1, 4), W (⁻3, 2)
  (b) Q (4, ⁻1), R (3, ⁻5), S (⁻2, ⁻4), T (⁻3, ⁻1)

Ask your teacher what to do next.

Keep it simple

Sometimes you can solve a **difficult** problem by
* first solving a **simpler** one of the same type
* then using the same method for the difficult problem.

In questions **1** and **2**
* **read** the problem in part **(a)**
* answer the **simpler** problem in part **(b)**
* go back and answer the problem in part **(a)**
  using the same method.

1 (a) | Problem |

Light travels at a speed of 310 000 km per second.
The distance from the Sun to the Earth is 155 000 000 km.
How long does it take light to travel from the Sun to the Earth?

(b) | Simpler problem |

Jim cycles to a friend's house at a speed of 10 km per hour.
The distance between their houses is 30 km.
* How long does Jim's journey take?
* Explain how you worked out your answer.

2 (a) | Problem |

In his shop last week Patrick's takings came to a total of £12 560.
He made a profit of £3 454.
Calculate his profit as a percentage of his takings.

(b) | Simpler problem |

Bill sold his personal stereo for £20.
He made a profit of £7.
* Calculate his profit as a percentage of the selling price.
* Explain how you worked out your answer.

Finding solutions to **several** simpler cases often helps to
solve a difficult problem. By listing the results systematically
you may see a pattern which leads you to the answer.

3 Each student in a class of 20 gives a Christmas card to every
other student. How many cards are given?

First find the answers for 1, 2, 3, 4 and 5 students in the class.
List the results in a table like this

Number of students	1	2	3	4	5		20
Number of cards	0						

4 There are 27 girls in Morven's guide company. They all have
lots of comics and decide to exchange them with one
another. Each girl gives all the other girls a comic.
How many comics are exchanged?

5

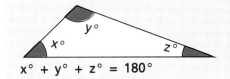

$x° + y° + z° = 180°$

The sum of the angles in a
triangle is **2 right angles.**

Remember

A **nonagon** is a shape with 9 sides.
How many right angles are there in the sum of its angles?

Try some simpler cases, for example,
a quadrilateral.

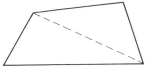

A table like this may help.

Number of sides in the shape	3	4	5			9
Number of right angles	2					

6 A group of **ten** friends all wanted to
play on the see-saw in the park. They
decided that, to be fair, each of them
would partner each of the others for
one turn. How many different pairs
were there?
First try working out the number of
pairs for smaller groups.

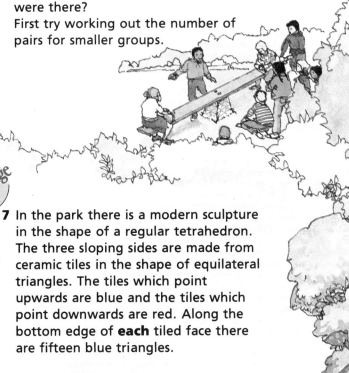

Challenge

7 In the park there is a modern sculpture
in the shape of a regular tetrahedron.
The three sloping sides are made from
ceramic tiles in the shape of equilateral
triangles. The tiles which point
upwards are blue and the tiles which
point downwards are red. Along the
bottom edge of **each** tiled face there
are fifteen blue triangles.

Find the total number of tiles on the sculpture.

A **conjecture** is a reasonable guess based on all the available evidence.

To **test** a conjecture more evidence has to be collected and the conjecture checked to see if it is still true when the new data is added.

Week	Dry days	Wet Days
1	3	4
2	2	5
3	3	4
4	1	6

1 Class 3 at Strangehill has collected data about the weather. They put their data in this table.

(a) From this data they made the conjecture:
In any week there are more wet days than dry days.
Does their conjecture fit with the data they collected?

(b) The class gathered data for another 3 weeks. Does their conjecture still hold when the new data is added? Explain.

Week	Dry Days	Wet Days
5	4	3
6	2	5
7	6	1

2 Jean and Sajidah measure the length of an elastic band when different weights are hung from it.
Sajidah draws a graph of their results.

(a) **Do Workbook page 39, question 1.**

(b) Jean looks at the graph and makes the conjecture:
The graph of length against weight is a straight line.
Does Jean's conjecture fit with their results?

(c) The girls repeat the experiment with other weights and obtain the results given in this table.
Does Jean's conjecture still hold?

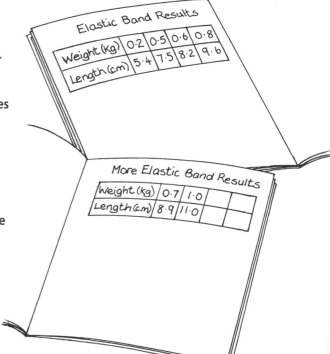

Elastic Band Results

Weight (kg)	0.2	0.5	0.6	0.8
Length (cm)	5.4	7.5	8.2	9.6

More Elastic Band Results

Weight (kg)	0.7	1.0		
Length (cm)	8.9	11.0		

3 Robin makes a list of the factors of all the numbers from 1 to 20.

(a) **Do Workbook page 39, question 2.**

(b) Robin makes the conjecture:
All prime numbers have exactly 2 factors.
Do you agree?   (Remember: 1 is **not** a prime number.)

(c) Test his conjecture for 3 prime numbers which are **not** in the table.

(d) Make a conjecture about the type of numbers which have an odd number of factors.

(e) Test your conjecture by using numbers which are not in the table.

	Product	Sum of digits
9 × 1	9	9 = 9
9 × 2	18	1 + 8 = 9
9 × 3	27	2 + 7 =
9 × 4	36	
9 × 5	45	
9 × 6	54	
9 × 7	63	
9 × 8	72	
9 × 9	81	

4 Lee noticed a strange fact about the products in the nine times table.

(a) Add the digits in each product in the table.
Make a conjecture about the sum of the digits of numbers which are exactly divisible by 9.

(b) Test your conjecture by using the next 3 numbers exactly divisible by 9.

(c) Alter your conjecture so that it is true for these new numbers as well.

(d) Test your new conjecture.

When Lesley went into her Maths Classroom she saw that the teacher had not cleaned the blackboard.

$$25 \xrightarrow[\text{THE DIGITS}]{\text{REVERSE}} 52 \quad \text{SUBTRACT THE SMALLER NUMBER FROM THE LARGER} \quad \begin{array}{r} 52 \\ -25 \\ \hline 27 \end{array}$$

$$64 \longrightarrow 46 \quad \begin{array}{r} 64 \\ -46 \\ \hline 18 \end{array}$$

That looks interesting. I'll try it with some other   numbers.

Wow!   It might be...

Eureka! All the answers..

5 (a) Copy and complete Lesley's conjecture.

(b) Test the conjecture.

Challenge

SNAP-A-PIC ORDERS

1	2	3				17	18
19	20	21					
						35	36

1 Vicky works in the Snap-a-pic photography shop. She sends negatives for printing. Each negative has a number. Negatives with the same number must be put in separate envelopes.

How many envelopes does Vicky need for each set of negatives?

(a)

(b)

2 A frame must allow for at least a 2 cm margin around all sides of a photograph.

  (a) Select the smallest possible frame for a photograph of size • 16 cm × 10 cm • 20 cm × 18 cm.

  (b) What is the largest possible photograph which fits each size of frame?

  (c) For any size of frame, write a rule for finding the dimensions of the largest possible photograph.

Frame sizes in cm

Small 16×12
Medium 26×20
Large 32×22

3 Time allowed for developing a film is **3 working days,** that is not counting Saturday or Sunday.

For films handed in after 2 pm time starts from the next day.
On what day should Andy's film be ready if he hands it in on

  (a) Monday 20 September at 9 am

  (b) Friday 24 September at 1 pm

  (c) Wednesday 29 September at 2.30 pm?

4 Alison is re-tiling the counter which measures 1 m by 2·4 m.
Each title is 20 cm square.

  (a) How many tiles does she need?

  (b) She tiles the counter in the pattern shown.
  How many tiles of each colour does she use?

The Friends of Queen Margaret Hospice held a fund-raising quiz night. They sold raffle tickets with 3-digit numbers.

1 (a) Marcia won the star prize. The three digits on her winning ticket totalled 27. What was the number on her ticket?

(b) The other prizes went to the holders of tickets with digits totalling *Lucky 21* .

What were the numbers on these tickets?

759 *Lucky 21*

GÂTEAU 50P

ÉCLAIR 40P

CHEESE SANDWICH 20P

HAM SANDWICH 25P

2 (a) During the interval the Friends sold cakes and sandwiches and raised a total of £240. The income from the cakes was **double** the income from the sandwiches. How much money was raised by selling
  • cakes   • sandwiches?

(b) The income from the gâteaux was **three times** the income from the éclairs.
How many cakes of each type were sold?

(c) The income from the ham sandwiches was £8 **more** than the income from the cheese sandwiches.
How many sandwiches of each type were sold?

3 Each team in the quiz sat at a numbered table. The players wore rosettes in their team's colours. The top four prizes were won by the teams at tables 23, 49, 56 and 78.
Use the clues to find • each team's colour
  • the prize won by the team at each table.

Challenge

The team at table 23 won the next prize **after** the team with blue rosettes.

The team with red rosettes won first prize.

The team which won second prize wore blue rosettes.

The team at table 78 wore pink rosettes.

Ask your teacher what to do next.

When added, the table numbers of the team with orange rosettes and the team which won second prize gave an even number.

The Glacier Valley Hotel

The Glacier Valley Hotel employs foreign students during the summer.

1 Hans works in the kitchen. He can fillet a fish in 10 minutes.
Copy and complete this table.

Number of fish	1	2	3	4	6
Time in minutes					

2 What happens to the time taken when the number of fish he fillets is

(a) doubled **(b)** trebled
(c) multiplied by 4 ?

3 Copy and complete:
When the time increases the number of fish ☐ in proportion.

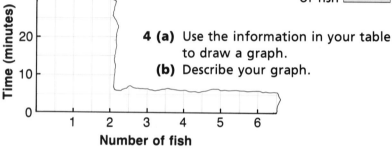

4 (a) Use the information in your table to draw a graph.
(b) Describe your graph.

The time taken and the number of fish filleted are in **direct proportion**.
The graph is a **straight line** passing through the origin.

5 Heidi works in reception. Copy and complete this table using the graph.

Hours worked	40	30	20	10
Pay (£)				

6 What happens to Heidi's pay when the number of hours she works is

(a) halved **(b)** quartered
(c) divided by 3 ?

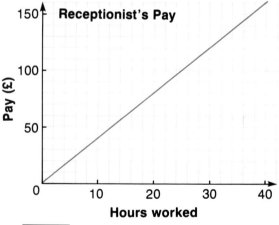

7 Copy and complete:

This graph is a ☐ passing through the ☐.
The hours worked and the pay earned are in ☐.

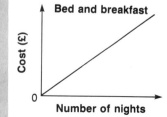

8 The graph shows the cost of B & B at the Glacier Valley Hotel.
(a) Describe the graph.
(b) Copy and complete:
The cost and the number of nights are in ☐.
(c) What happens to the cost when the number of nights
• doubles • halves • trebles • quarters
• is multiplied by 5 • is divided by 6?

The Triple Star Club are off on a Continental tour. Each member has a supply of travellers cheques. Jo the leader bought some foreign currency.

Country	Exchange rate per £
Austria	20·25 schillings
Belgium	59·30 francs
France	9·80 francs
Italy	2170 lira

At the bank Jo exchanged £25 for Belgian francs. How many did she receive?

£	Francs
1	59·30
25	59·30 × 25 = 1482·50

Jo received **1482·50 francs.**

1 Calculate the amount of foreign currency Jo received when she exchanged
   (a) £100 for Belgian francs   **(b)** £50 for schillings
   (c) £75 for lira   **(d)** £45 for French francs.

At the German hotel Terry exchanged £7 and received 18·30 deutchmarks. How many deutchmarks would Val receive for £12 ?

£	Dm
7	18·30
1	18·30 ÷ 7
12	18·30 ÷ 7 × 12

Enter `18.30` Press `÷` `7` `×` `1` `2` to give `31.371428`

Val would receive **31·37 Dm**.

2 At the hotel how many deutchmarks would be exchanged for
   (a) £5   **(b)** £9   **(c)** £17   **(d)** £24 ?

3 At the hostel in Switzerland Anya exchanged £15 for 36·75 francs.
   How many Swiss francs would be exchanged for
   (a) £10   **(b)** £4   **(c)** £25   **(d)** £70?

4 Copy and complete the table to find the cost of Jo's Paris fashion scarf in British money.

Francs	£
9·80	1
1	1 ÷ 9·80
86	

5 Find the cost in British money of each of the following.
   (a) Wallet from Austria   **(b)** Necklace from Italy   **(c)** Sunglasses

AS 177

L 15 000

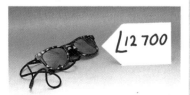

L 12 700

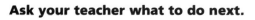

Ask your teacher what to do next.

PART
FOUR

Ian and John wanted to see if there was a connection between the **circumference** and the **diameter** of a circle.

They measured the diameter and the circumference of the top of a soup tin.

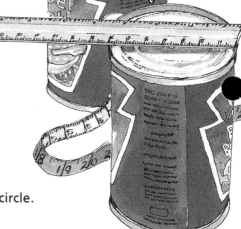

You need several shapes with circular ends.

1 Measure the diameter and circumference of each circle.
Put your results in a table like this:

Shape	diameter, d	Circumference, C
Soup tin	7 cm	22 cm
Waste paper bin		

2 Copy and complete:
For a circle, the circumference is about ☐ times the diameter.

3 Use the formula $C = 3d$ to find the approximate circumference of:
   (a) the button   **(b)** the plate   **(c)** the tyre   **(d)** the compact disc.

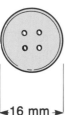
◄16 mm►

◄— 27 cm —►

◄— 32 mm —►

◄— 12 cm —►

4 Measure the diameter in millimetres and calculate the approximate circumference of each coin.
   (a)   **(b)**   **(c)**   **(d)**   **(e)**

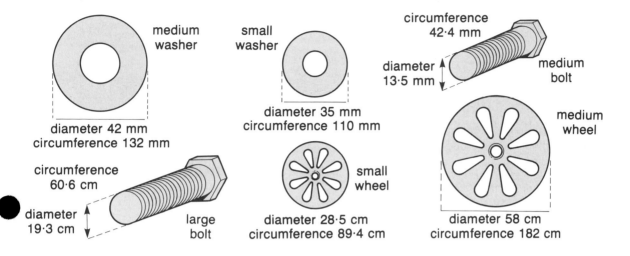

medium washer

diameter 42 mm
circumference 132 mm

small washer

diameter 35 mm
circumference 110 mm

circumference 42·4 mm

diameter 13·5 mm

medium bolt

medium wheel

circumference 60·6 cm

diameter 19·3 cm

large bolt

small wheel

diameter 28·5 cm
circumference 89·4 cm

diameter 58 cm
circumference 182 cm

Product	Circumference C	diameter d	Circumference ÷ diameter, $\frac{C}{d}$ calculator answer	$\frac{C}{d}$ rounded to 2 decimal places
medium washer small washer	132 mm	42 mm	3·1428571	3·14

We already know that the circumference is about 3 times the diameter.
It would be **more accurate** to say that
the circumference is **3·141592653589** . . . . . times the diameter.
This number is represented by the Greek letter π, called **pi**.

To calculate the circumference of a circle we use the formula
$C = \pi d$ with π = 3·14 rounded to two decimal places.

2 In the past there have been different values for pi.
Which value of π was most accurate?

$\pi = \frac{25}{8}$

2000 BC BABYLON

$\pi = \frac{256}{81}$

1600 BC EYGPT

$\pi = \frac{377}{120}$

150 AD GREECE

$\pi = \frac{355}{113}$

480 AD CHINA

The diameter of a wheel is 7 cm.
Calculate its circumference.

$C = \pi d$
$C = 3 \cdot 14 \times 7$
$C = 21 \cdot 98$

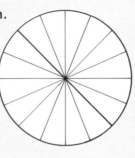

The circumference is **21·98 cm**,
to two decimal places.

1 Use the formula $C = \pi d$ to calculate the
circumference of each of these wheels.

(a)

diameter 28 cm

(b)

diameter 8 cm

(c)

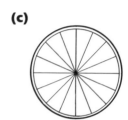

diameter 15·1 cm

2 For each toy car wheel, measure the diameter
and then calculate its circumference.

(a)

(b)

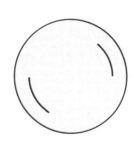

(c)

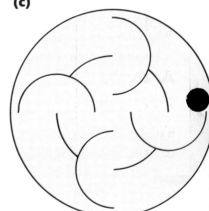

3 The diameter of a bicycle wheel is 60 cm.
Calculate its circumference in **metres.**

Challenge

4 This tractor has large wheels of diameter 1·5 m
and small wheels of diameter 0·5 m. Calculate
  (a) the circumference of a large wheel
  (b) the circumference of a small wheel
  (c) how far the tractor travels in 30
     revolutions of the large wheel
  (d) the number of revolutions made by the small
     wheel over the same distance.

Ask your teacher what to do next.

Class 3 at Strangehill School is collecting 20p coins for a local charity. They have chosen different fund raising activities.

Work in a group.

For each activity
- discuss how to solve the problem
- list the equipment you will need
- find how much money they should collect.

Activity 1

The 10 metre length

Lay out 20p coins to make a 10 m line.

Activity 2

The metre tube

Fill the tube with 20p coins.

Activity 3

The cola can tower

Build a tower of 20p coins around and up to the top of a soft drinks can.

Activity 4

The 20p kilogram

Balance the scales – fill the bag with 20p coins.

Activity 5

The square metre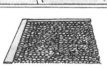

Cover the square metre with 20p coins.

Ask your teacher what to do next.

You need scissors, glue and card.

For questions **1** and **2** you **must** work with a partner.

1 Cut out Workbook page 24.
- Cut out the two nets.
- Glue the nets to card and cut them out.
- One of you colour **one** of the faces marked (A) and the other colour **one** of the faces marked (B).
- Fold along the dotted lines and glue down the flaps. The coloured faces should be on the outside.

You and your partner should now have four solids. Two of the solids should have one coloured face each.

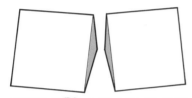

- Stick the coloured faces together with a piece of card between them.

The triangular prism you have made is **balanced** about the joining card. The joining card is a **plane of symmetry.**

2 • Glue a third solid on the end to make an **unbalanced** triangular prism.
- Re-balance the prism with your fourth solid.

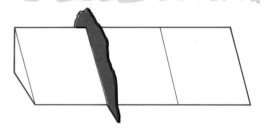

3 • Make enlargements of these nets by drawing them on 1 cm squared paper. (Include flaps where you think you will need them.)
- Glue the nets to card and cut them out.
- Stick the coloured faces together to make a single 3D-shape with a plane of symmetry.

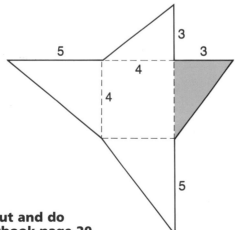

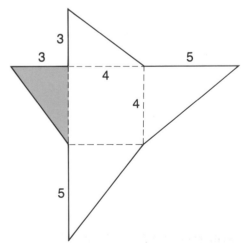

4 Cut out and do Workbook page 20.

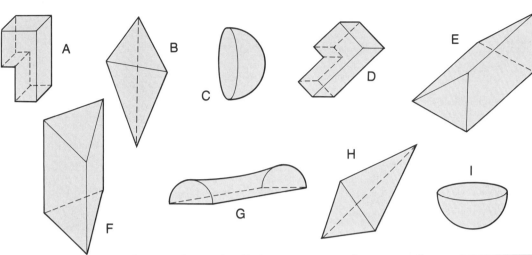

1 You can join pairs of these shapes by fitting congruent faces together so that the resulting solids have a plane of symmetry. One pair has been joined for you.
Which other pairs of shapes can be joined in this way?

2 Jill has put up some decorations for her party. Which pictures show a decoration cut by a plane of symmetry?

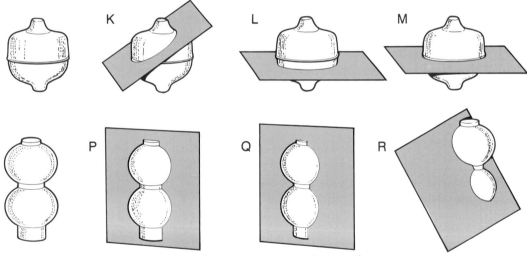

3 Jill received presents in these boxes. How many planes of symmetry are there for each box?

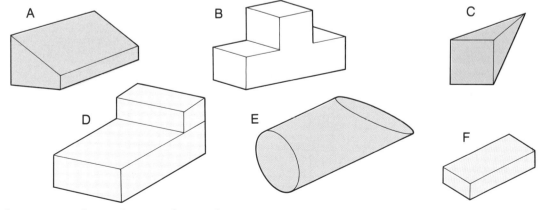

Ask your teacher what to do next.

The Labyrinth

Rona has designed an adventure game, The Labyrinth, where players find their way through a maze of rooms.

The Labyrinth has 5 rooms. Each room has a number of doors of different colours.

Lindsay is going to play the game.
She will be placed at random in one of the five rooms.

The probability that she will start in the Dungeon is $\frac{1}{5}$.
This can be written as

P (Dungeon) = $\frac{1}{5}$

1 For each new game that Lindsay starts find

(a) P (Captain's room)
(b) P (Treasury)
(c) P (room with 3 doors)
(d) P (room with 4 doors)
(e) P (room with 5 doors)
(f) P (room with a blue door)
(g) P (room with a green door)
(h) P (room with a red door)
(i) P (room with a blue door **and** a yellow door)
(j) P (room with a blue door **or** a yellow door).

2 Talat starts the game in the **Creepy room**.
He selects a door at random. Find

(a) P (red door)
(b) P (door is not red)
(c) P (green door)
(d) P (yellow or red door).

3 Talat enters the **Dungeon** where there are 12 levers.

> 1 lever opens only the red door,
> 5 levers open only the blue door,
> 4 levers open all the doors and
> 2 levers lock all the doors FOREVER.

Talat selects a lever at random.
Find
(a) P (only the blue door opens)
(b) P (the blue door opens)
(c) P (only the red door opens)
(d) P (the red door opens)
(e) P (the yellow door opens)
(f) P (the green door opens)
(g) P (all the doors open)
(h) P (none of the doors opens).

Blair has 3 shirts: blue (b), white (w), grey (g)
and 4 ties: red (r), blue (b), yellow (y), grey (g).

1 (a) Copy and complete the tree
diagram to show all the possible
choices of shirts and ties that Blair
can wear.

(b) List all the possible outcomes like
this:
br, bb, by, ...

(c) Copy and complete:
Number of shirts   = ☐
Number of ties   = ☐
Number of choices = ☐

Shirts   Ties

b ⟨ r b y g

w ⟨ r b

There are **12** possible outcomes.
so **P (white shirt and red tie)** = $\frac{1}{12}$

2 Blair chooses a shirt and a tie at random. Find

(a) P (blue shirt and red tie)   **(b)** P (blue shirt and blue tie)
(c) P (same colour shirt and tie) **(d)** P (different colour shirt and tie).

3 Blair also has 2 jackets: blue (b), checked (c)
and 5 pairs of trousers: blue (b), grey (g),
striped (s), checked (c), white (w).

(a) Copy and complete the tree
diagram to show all the possible
choices of jackets and trousers.

(b) List all the possible outcomes.

(c) Copy and complete:
Number of jackets   = ☐
Number of pairs of trousers = ☐
Number of choices   = ☐

Jackets   Trousers

b ⟨ b g s c w

4 Blair chooses a jacket and a pair of trousers at random. Find

(a) P (blue jacket and grey trousers) **(b)** P (blue jacket and blue trousers)
(c) P (similar jacket and trousers)   **(d)** P (different jacket and trousers)
(e) P (white trousers)   **(f)** P (checked jacket)

5 Make a tree diagram and list all the
possible ways that Blair can choose a
shirt, a tie and a jacket.

6 Blair chooses a shirt, a tie and a jacket
at random. Find

(a) P (white shirt, red tie and blue jacket)
(b) P (similar shirt, tie and jacket)
(c) P (similar shirt and tie)
(d) P (similar tie and jacket)

Shirts   Ties   Jackets

b ⟨ r b y g → b c b c b c

7 What is the probability of Blair choosing at random, a blue
shirt, a blue tie, a blue jacket and blue trousers?

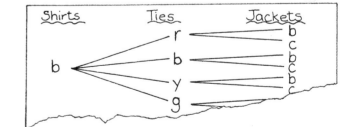

SATURDAY SUPERSHOW

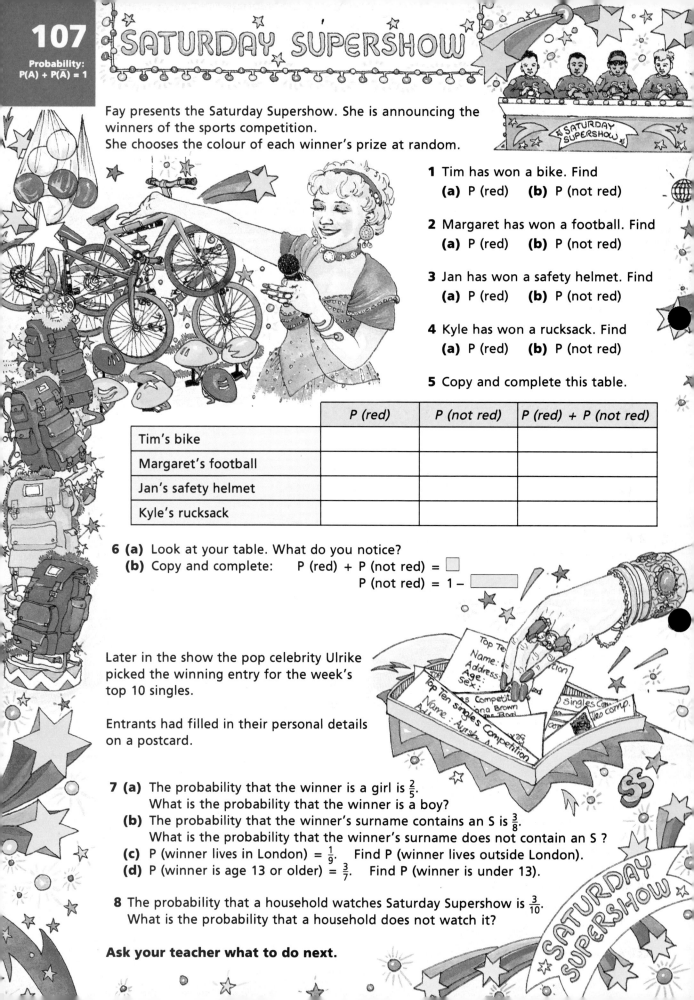

Fay presents the Saturday Supershow. She is announcing the winners of the sports competition.
She chooses the colour of each winner's prize at random.

1 Tim has won a bike. Find
  (a) P (red)   **(b)** P (not red)

2 Margaret has won a football. Find
  (a) P (red)   **(b)** P (not red)

3 Jan has won a safety helmet. Find
  (a) P (red)   **(b)** P (not red)

4 Kyle has won a rucksack. Find
  (a) P (red)   **(b)** P (not red)

5 Copy and complete this table.

	P (red)	P (not red)	P (red) + P (not red)
Tim's bike			
Margaret's football			
Jan's safety helmet			
Kyle's rucksack			

6 (a) Look at your table. What do you notice?
  (b) Copy and complete: P (red) + P (not red) = ☐
                                P (not red) = 1 – ☐

Later in the show the pop celebrity Ulrike picked the winning entry for the week's top 10 singles.

Entrants had filled in their personal details on a postcard.

7 (a) The probability that the winner is a girl is $\frac{2}{5}$.
    What is the probability that the winner is a boy?
  (b) The probability that the winner's surname contains an S is $\frac{3}{8}$.
    What is the probability that the winner's surname does not contain an S ?
  (c) P (winner lives in London) = $\frac{1}{9}$.   Find P (winner lives outside London).
  (d) P (winner is age 13 or older) = $\frac{3}{7}$.   Find P (winner is under 13).

8 The probability that a household watches Saturday Supershow is $\frac{3}{10}$.
   What is the probability that a household does not watch it?

Ask your teacher what to do next.

Toddler's Togs

🅣 Aberdeen Branch 🅣
Manager : Duncan Grant

Sales reps : Anne Watson
David Ingram
Morag MacDonald

  SALES TEAMS

🅣 Newcastle Branch 🅣
Manager: Ahmed Baruz

Sales reps : Doug Thomson
Amanda Collins
Sue Drummond

🅣 Cardiff Branch 🅣
Manager : Bethan Davies

Sales reps : Peter Schuman
Tom Collins

🅣 London Branch
Manager : Lisa Gibling

Sales rep : June Davidson

The sales teams for Toddler's Togs travel between branches.
The return air fares are given below.

Aberdeen

£170	Newcastle		
£230	£155	Cardiff	
£275	£175	£110	London

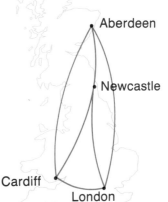

1 Find the cost of each return journey.

  (a) Bethan travels to Newcastle to meet a client.

  (b) Morag travels to London to deliver designs.

  (c) Doug travels to Aberdeen for a meeting.

2 Find the cost for all 4 managers to
travel to a convention in Newcastle.

3 The managers have regular meetings
to discuss company plans.
In which of the 4 branches should they
meet to keep the cost of air fares as
low as possible?

4 In June the sales teams attend their annual conference.
Which branch should host the conference to keep the air
fares as low as possible?

5 The sales teams attend a 2-day
training course in November. This
involves **everyone** staying overnight
in a hotel.
The costs are shown opposite.
Where should they meet to keep the
overall cost as low as possible?

6 What else could be taken into account
when deciding where to hold a
meeting?

Ask your teacher what to do next.

In the pond there are 5 ducks and 4 geese.
The **ratio** of **ducks to geese** is **5 to 4**
written as **5:4**

The ratio of **geese to ducks** is **4 to 5**
or **4:5**.

1 Write the ratio of

(a) eggs to hens
(b) hens to eggs

(c) sheep to lambs
(d) lambs to sheep

(e) sheep to dogs
(f) dogs to sheep

(g) tractors to trailers
(h) trailers to tractors.

2 Horatio plants 3 fields with oats and 5 fields with barley.
Write the ratio of **(a)** oats fields to barley fields
(b) barley fields to oats fields.

3 In a field there are 15 cows and 22 calves.
Write the ratio of **(a)** calves to cows
(b) cows to calves.

4 For his tea Horatio has 2 eggs,
3 sausages, and 5 mushrooms.
The ratio of eggs to mushrooms is 2:5.
Write the ratio of **(a)** eggs to sausages
(b) mushrooms to sausages.

5 On Jenny's plate what is the ratio of

(a) sausages to eggs
(b) tomatoes to sausages
(c) eggs to tomatoes?

6 On the table there are 7 scones, 3 cakes, and 4 rolls.
Write the ratio of **(a)** scones to cakes **(b)** rolls to scones **(c)** cakes to rolls.

In one hen house 8 hens have laid a total of 28 eggs.

The ratio of eggs to hens is **28:8**.

The eggs and hens can be arranged in 4 equal groups.

The ratio of eggs to hens is **7:2**.

> **The ratio 28:8 in simplest form is 7:2.**

1 In the next hen house 9 hens have laid 24 eggs.
The hens and eggs are arranged in 3 equal groups.

(a) How many • hens • eggs   are in each group?
(b) Write the ratio of hens to eggs in simplest form.
(c) Copy and complete:
The ratio 9:24 in simplest form is ☐.

2 Horatio uses 4 dogs to look after his 100 sheep.
The dogs and sheep are arranged in 4 equal groups.

(a) Write the ratio of dogs to sheep in simplest form.
(b) Copy and complete:
The ratio 4:100 in simplest form is ☐.

3 Horatio's herd of 24 cows has given birth to 32 calves.
The cows and calves are arranged in 8 equal groups.

(a) Write the ratio of cows to calves in simplest form.
(b) Copy and complete:
The ratio 24:32 in simplest form is ☐.

4 Horatio has 10 pigs and 105 piglets.

(a) What is the maximum number of equal groups of
pigs and piglets?
(b) Write the ratio of piglets to pigs in simplest form.
(c) Copy and complete:
The ratio 105:10 in simplest form is ☐.

5 Write each ratio in simplest form.

(a)	4:10	**(b)**	6:15	**(c)**	5:35	**(d)**	12:28	**(e)**	6:6
(f)	18:4	**(g)**	9:6	**(h)**	36:16	**(i)**	28:12	**(j)**	56:21
(k)	14:35	**(l)**	35:15	**(m)**	12:42	**(n)**	40:50	**(o)**	45:15

The Citron Car Company

The Citron Car Company makes cars in kit form.
Liam paints the cars either yellow or green.
The ratio of yellow to green cars is **1:3**.

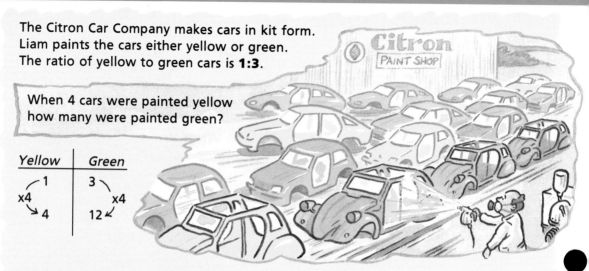

When 4 cars were painted yellow
how many were painted green?

Yellow	Green
⌐1	3⌐
×4	×4
↘4	12↙

There were **12** green cars.

1 Copy and complete the tables.

(a)

Yellow	Green
⌐1	3⌐
×5	×5
↘5	☐↙

(b)

Yellow	Green
⌐1	3
×9	⌐
↘9	☐

(c)

Yellow	Green
⌐1	3⌐
	×6
☐	18↙

(d)

Yellow	Green
1	3
↓	⌐
☐	60

2 (a) How many cars were painted green when 30 were painted yellow?
   (b) How many cars were painted yellow when 30 were painted green?

The Citron cars use either petrol or diesel.
The ratio of petrol to diesel is 5:2

How many diesel cars were made when
the number of petrol cars was 20 ?

Petrol	Diesel
⌐5	2⌐
×4	×4
↘20	8↙

There were **8** diesel cars made.

3 How many diesel cars were made when the number of petrol cars was
   (a) 10   **(b)** 25   **(c)** 35   **(d)** 50   **(e)** 75 ?

4 How many petrol cars were made when the number of diesel cars was
   (a) 6   **(b)** 16   **(c)** 18   **(d)** 22   **(e)** 40 ?

5 How many left hand drive cars were
   made when the number of right hand
   drive cars was
   (a) 6   **(b)** 15   **(c)** 24   **(d)** 30 ?

6 How many right hand drive cars were
   made when the number of left hand
   drive cars was
   (a) 24   **(b)** 32   **(c)** 56   **(d)** 72 ?

Car Exports
Citron

RIGHT HAND . LEFT HAND
DRIVE   DRIVE

3 : 8

Ask your teacher what to do next.

Philip works for Wonderdrug, the
chemist. He uses a counter in the
shape of an equilateral triangle to
count the pills.

1 The diagrams show a pill counter with different numbers of rows filled.

 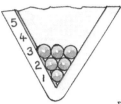

Draw a pill counter which is filled to
(a) the 4th row   **(b)** the 5th row.

2 Copy and complete the table. Use the pill counter to help.

Number of rows filled	1	2	3	4	5	6	7	8
Total number of pills			6					

3 Explain how you found the total number of pills in the counter
when it was filled to the 8th row.

4 What is the total number of pills in the counter when it is filled to
(a) the 9th row   **(b)** the 10th row?

1, 3, 6, 10, ... are known as **triangular numbers.**

5 You can show triangular numbers like this.
In the same way draw the next three
triangular numbers.

The 3rd triangular
number is 6.

1 + 2 + 3 = 6

The 4th triangular
number is 10.

1 + 2 + 3 + 4 = 10

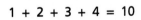

6 In the same way, write sums for   **(a)** the 5th triangular number
                                         (b) the 8th triangular number.

7 (a) Add   • the 1st and 2nd triangular numbers
                 • the 2nd and 3rd triangular numbers.
(b) Add other pairs of **consecutive** triangular numbers.
(c) Write about what you find.

Ask your teacher what to do next.

After school, Emma and Kandor fill and
weigh packets at the delicatessen,
Gourmet's Choice.

1 Write the weight in grams of each packet.

(a)

(b)

(c)

These scales are balanced.

Kandor takes 20 grams from each side.

The packet weighs **30 grams**.

2 Do Workbook page 40.

3 Emma solves this equation to find the weight, *w* grams, of the nutmeg.

Equation: $w + 30 = 80$

Take 30 from each side. $\boxed{-30}\ \boxed{-30}$

$w \qquad = 50$

The nutmeg weighs **50 grams**.

For each balance, copy and complete the equation. Solve it to find the weight in
grams of the bag.

(a)

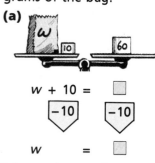

$w + 10 = \square$
$\boxed{-10}\ \boxed{-10}$
$w \quad = \square$

(b)

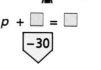

$p + \square = \square$
$\boxed{-30}$
$p \quad = \square$

(c)

$200 = \square + x$
$\boxed{}\ \boxed{}$
$\square = x$

4 Solve each equation.

  (a) $w + 5 = 40$ **(b)** $p + 40 = 100$ **(c)** $70 = x + 20$

  (d) $y + 65 = 85$ **(e)** $12 + t = 30$ **(f)** $120 = t + 15$

  (g) $6 + n = 14$ **(h)** $r + 170 = 195$ **(i)** $20 = 3 + p$

1 For each balance find the weight in grams of **one** packet.

(a)

(b)

(c)

Kandor solves this equation to find the weight, t grams, of one bag of bay leaves.

$t + t + t + t + t = 55$

$5t = 55$

$t = 11$

$t + t + t + t + t = 5t$
$5 \times ? = 55$
$5 \times 11 = 55$

Each bag weighs **11 grams**.

2 For each balance write an equation.
Solve it to find the weight in grams of **one** bag.

(a)

(b)
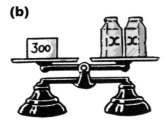

(c)

3 Solve each equation.

(a) $t + t + t = 60$ **(b)** $p + p + p + p = 100$ **(c)** $36 = x + x$
(d) $3y = 15$ **(e)** $4n = 40$ **(f)** $80 = 5m$
(g) $2t = 280$ **(h)** $750 = 5r$ **(i)** $8p = 96$

Emma solves this equation to find the weight, b grams, of one box of cheese.

Take 50 from each side.

$2b + 50 = 350$

$\boxed{-50}\ \boxed{-50}$

$2b = 300$
$b = 150$

Each box weighs **150 grams**.

4 Do Workbook page 41.

5 Solve each equation.

(a) $2x + 10 = 40$ **(b)** $3p + 15 = 45$ **(c)** $4 + 7b = 25$
(d) $26 = 4y + 6$ **(e)** $12 + 5n = 17$ **(f)** $1 + 2x = 7$
(g) $3m + 5 = 11$ **(h)** $4r + 30 = 130$ **(i)** $31 = 6 + 5v$

Andrew is stocktaking at Kay's the chemist.

He finds the weight in kilograms of **one** of these bottles by solving this equation:

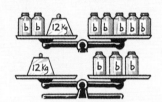

Take 2b from each side.

Each bottle weighs **4 kg.**

$$2b + 12 = 5b$$
$$-2b \qquad -2b$$
$$12 = 3b$$
$$4 = b$$

1 Do Workbook page 42.

2 Solve each equation.

  (a) $2b + 15 = 5b$ **(b)** $x + 18 = 7x$ **(c)** $9y = 4y + 10$
  (d) $36 + 2n = 8n$ **(e)** $5 + 3p = 8p$ **(f)** $6x = 2x + 28$

Andrew solves this equation to find the weight in kilograms of one jar.

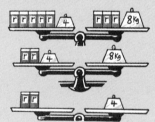

Take 3r from each side.

Take 4 from each side.

Each jar weighs **2 kg.**

$$5r + 4 = 3r + 8$$
$$-3r \qquad -3r$$
$$2r + 4 = \qquad 8$$
$$-4 \qquad -4$$
$$2r = 4$$
$$r = 2$$

3 Do Workbook page 43.

4 Solve each equation.

  (a) $3x + 1 = 2x + 6$ **(b)** $4y + 2 = y + 14$ **(c)** $2n + 20 = 5n + 2$
  (d) $5r + 2 = 3r + 8$ **(e)** $7k + 19 = 8k + 10$ **(f)** $t + 20 = 4t + 5$
  (g) $5 + 4b = b + 17$ **(h)** $6m + 1 = 11 + m$ **(i)** $15 + 3x = 3 + 7x$

5 Solve each equation.

  (a) $p + 4 = 7$ **(b)** $5 + x = 9$ **(c)** $16 = h + 7$
  (d) $4c = 24$ **(e)** $32 = 8p$ **(f)** $2x + 6 = 10$
  (g) $15 = 3v + 6$ **(h)** $4 + 3v = 19$ **(i)** $25 = 1 + 3t$
  (j) $3x + 20 = 5x$ **(k)** $8 + p = 5p$ **(l)** $5y = 2y + 51$
  (m) $4a + 3 = a + 9$ **(n)** $20 + 3v = 8v + 5$ **(o)** $5 + 6k = 10k + 1$

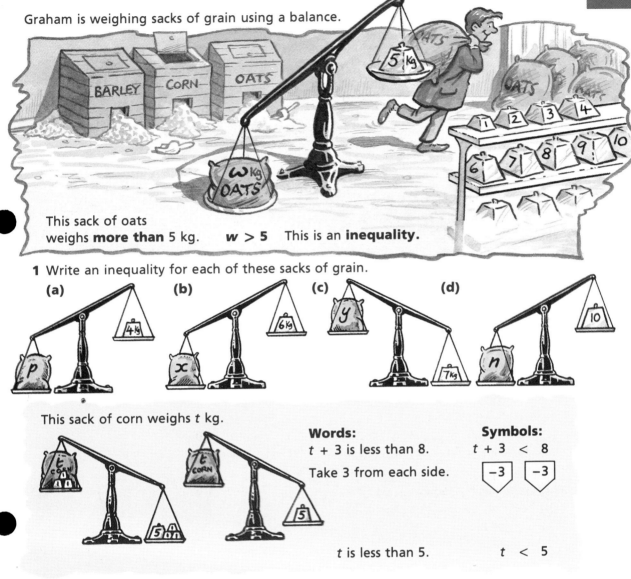

Graham is weighing sacks of grain using a balance.

This sack of oats
weighs **more than** 5 kg. $w > 5$ This is an **inequality**.

1 Write an inequality for each of these sacks of grain.

(a)   **(b)**   **(c)**   **(d)**

This sack of corn weighs t kg.

Words:

$t + 3$ is less than 8.

Take 3 from each side.

t is less than 5.

Symbols:

$t + 3 < 8$

$\boxed{-3}$ $\boxed{-3}$

$t < 5$

2 For each of these sacks, use **symbols** to write an inequality.
Find its solution.

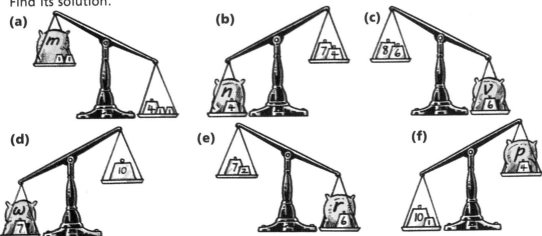

(a)   **(b)**   **(c)**

(d)   **(e)**   **(f)**

Ask your teacher what to do next.

Starship Endeavour

We're preparing the Starship Endeavour for a five-year mission.

Captain Cosmos

1st Officer Mork

Lt. Daz

Dr. Splint

Zark

Lt. Pravchek

Major Day

Paul

PLANET EARTH: STARDATE 2344.

1 This is the logo of the Starship Endeavour. On $\frac{1}{2}$ cm squared paper draw the logo when it is

(a) enlarged by a scale factor of 2

(b) reduced by a scale factor of $\frac{1}{2}$.

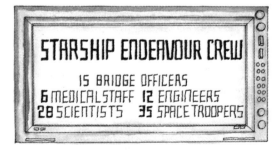

STARSHIP ENDEAVOUR CREW

15 BRIDGE OFFICERS
6 MEDICAL STAFF   12 ENGINEERS
28 SCIENTISTS   35 SPACE TROOPERS

2 Write, **in simplest form,** the ratio of

(a) engineers to medical staff

(b) bridge officers to space troopers

(c) scientists to engineers

(d) medical staff to scientists

(e) space troopers to scientists.

3 Each member of Endeavour's crew wears a hat with a green band. Find the length of band needed for a hat with diameter.

(a) 17 cm   (b) 18 cm   (c) 19 cm.

118

Extended
context

4

Dr. Splint
orders our
medical supplies.

SPACE TRAVELLERS MEDICAL CATALOGUE

■ SUNSPOT CREAM
Available sizes
3 litres
5 litres

■ TUMMY TAMERS
Available sizes
3 kg
7 kg

■ FAST-EAZ
Available sizes
200 pills
500 pills

■ MEDI-TAPE
Available sizes
5 metres
12 metres

Find the number and size of each
container he needs for exactly

(a) 17 litres of Sun Spot cream

(b) 32 kg of Tummy Tamers

(c) 56 metres of Medi-tape

(d) 2300 Fast-Eaz pills.

Lt. Pravchek, the Navigation Officer enters
into the ship's computer the chart references
for the planets and space stations the
Endeavour will visit during the mission.

- Station Zed, 28
- Station Omega, 7
- Vog, 37
- Darko, 99
- Anzig, 19
- Cora, 25
- Nast, 14
- Station Beta, 52
- Bepo, 6
- Plink, 36
- Kantor, 44
- Vangor, 81
- Kreep, 12

5 (a) The chart reference for Vog is ($^-$5, 8, 37). Write the chart reference for each of the
other planets and space stations.

(b) Make a chart on 2 mm graph paper.
Plot the following planets and space stations.
- Zangor (6, 5, 43)
- Quizard (34, $^-$5, 17)
- Perseus (7, $^-$7, 91)
- Mobius (22, 8, 79)
- Lizarak ($^-$10, 13, 62)
- Yoko ($^-$35, 0, 9)
- Zeus IV (0, 14, 99)
- Orak ($^-$15, $^-$19, 6)
- Rast ($^-$27, 6, 45)

The first part of our mission is to deliver an inter-planetary transporter to Perseus. Zark and Paul will have to assemble the transporter from six sections.

1 Cut out the nets on **Workbook pages 18 and 26.**
Make a model of the transporter.

2 The transporter has one plane of symmetry. Paul is standing on a platform directly above the transporter. Zark is standing in front of it. On **Workbook page 44** complete the views of the transporter seen by Paul and Zark.

3

Lt. Pravchek estimates the time on each part of the journey. Take-off is scheduled for 13 October 2344. We are intending to spend five days on each planet or space station. Copy and complete our timetable.

| October 2344 | | | | | | | | | | | | | | | | | | | | November 2344 | | | | | | | | | | | | | | | | | | |
| 13 | 14 | 15 | 16 | 17 | 18 | 19 | 20 | 21 | 22 | 23 | 24 | 25 | 26 | 27 | 28 | 29 | 30 | 31 | 1 | 2 | 3 | 4 | 5 | 6 | 7 | 8 | 9 | 10 | 11 | 12 | 13 | 14 | 15 | 16 | 17 | 18 | 19 |

← Earth to Perseus ——————————— Rest —— Perseus to ———

FROM	TO	DEPART	FLIGHT DURATION	ARRIVE
Earth	Perseus	13 Oct 2344	20 days	02 Nov 2344
Perseus	Zeus IV	07 Nov 2344	12 days	
Zeus IV	Zangor		8 days	
Zangor	Quizard		18 days	
Quizard	Vog		7 days	
Vog	Kantor		37 days	

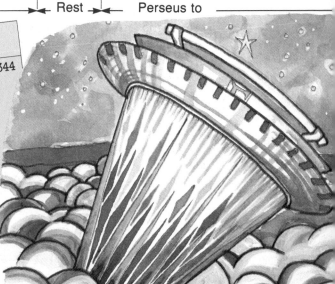

1 This graph shows the first four minutes of Endeavour's flight. What is the scale on
   (a) the horizontal axis   **(b)** the vertical axis?

2 At what height was Endeavour after
   (a) 3 min   **(b)** 1 min
   (c) 2·5 min   **(d)** 1·4 min?

3 How long did it take Endeavour to reach a height of
   (a) 3000 m   **(b)** 1600 m
   (c) 400 m   **(d)** 4400 m?

4 Find the average speed between 3 and 4 minutes after take-off.

5 After 4 minutes Endeavour continued to climb at this same speed. How many minutes after take-off did it take Endeavour to reach
   (a) 9000 m   **(b)** 11 000 m
   (c) 12 500 m   **(d)** 15 km?

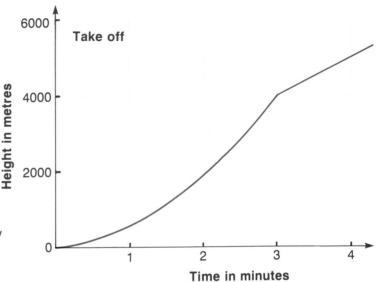

In space, a day is still the same length of time as on Earth but
   1 day   = 10 space hours (sh)
   1 sh   = 100 space minutes (sm)

The clock face shows Space time and Earth time.
   1200 Earth time = 5·0 sh Space time
   0600 Earth time = 2·5 sh Space time
   1900 Earth time = 7·9 sh Space time
      to the nearest 0·1 sh.

6 Write these Earth times as Space times to the nearest 0·1 sh.
   (a) 0300   **(b)** 0700   **(c)** 1400   **(d)** 1000
   (e) 2100   **(f)** 0500   **(g)** 1430   **(h)** 0615

7 **You need a pair of compasses, protractor and ruler.**
   Make an accurate drawing of the clock face.

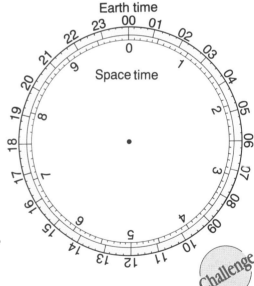

Challenge

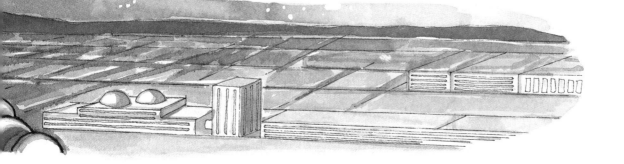

The Mobius Trip

Earth has lost contact with the Vogans on Space Station Mobius. From Perseus, the Endeavour is sent to investigate.

> I checked the Mobius floor plan. Unfortunately some of the computer's memory banks have been damaged and the data is incomplete.

The plan shows
- the Observation Room is located at (2, A4)
- the Main Power Room is located at (1, D1) and (1, E1).

MOBIUS FLOOR PLAN

LEVEL 2

	A	B	C	D	E
4	OBSERVATION ROOM	HOLD DECK	LIFT		
3			EVACUATION AREA		
2	RECREATION				WEAPONS
1			SHOWERS	GYMNASIUM	

1 What is located at
   (a) (2, B4)
   (b) (2, E2)
   (c) (1, E2)
   (d) (1, D3) and (1, E3)?

2 What is the location of
   (a) the Computer Room
   (b) the Emergency Power Room
   (c) the Recreation Area
   (d) the Lift?

LEVEL 1

	A	B	C	D	E
4	EMERGENCY POWER		LIFT	COMPUTER ROOM	
3				MEDICAL BAY	
2	TRANSPORTER ROOM				CAPTAIN'S ROOM
1			ENTRY AREA	MAIN POWER	

3 The Landing Party of Cosmos, Pravchek, Mork, and Zark arrive at the Entry Area. They gain entry by inserting an ID card and typing in a code based on their date of birth. The Data Sheet has an example of the code.

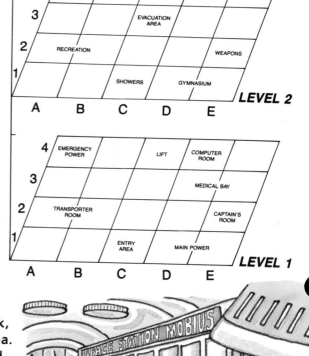

```
DATA SHEET 2        SPACE STATION MOBIUS
                    SECURITY ACCESS

DATE OF BIRTH       CODE
24/07/15            201475
```

(a) Use this information to write the access code for
   • Pravchek: born 23/09/98 • Mork: born 26 December 2306.
(b) What are the dates of birth of
   • Zark: access code 201029 • Cosmos: access code 110905.

4 First Officer Mork radios back the locations of the rooms on Mobius.
Complete the floor plan of Mobius on **Workbook page 45**.

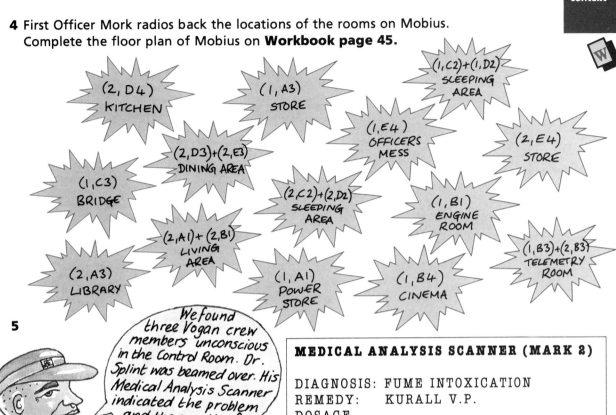

(2,D4) KITCHEN

(1,A3) STORE

(1,C2)+(1,D2) SLEEPING AREA

(2,D3)+(2,E3) DINING AREA

(1,E4) OFFICERS MESS

(2,E4) STORE

(1,C3) BRIDGE

(2,C2)+(2,D2) SLEEPING AREA

(1,B1) ENGINE ROOM

(2,A1)+(2,B1) LIVING AREA

(1,B3)+(2,B3) TELEMETRY ROOM

(2,A3) LIBRARY

(1,A1) POWER STORE

(1,B4) CINEMA

5

We found three Vogan crew members unconscious in the Control Room. Dr. Splint was beamed over. His Medical Analysis Scanner indicated the problem and the equation for the remedy.

MEDICAL ANALYSIS SCANNER (MARK 2)

DIAGNOSIS: FUME INTOXICATION
REMEDY: KURALL V.P.
DOSAGE
EQUATION: $3d+a=w$
d=dose (ml) a=age (years) w=weight (kg)

Solve the equation to find the dose, in millilitres, required for each Vogan.

(a)

VOGAL ID
Age: 30 years
Weight: 75 kg

(b)

VOGEB ID
Age: 25 years
Weight: 85 kg

(c)

VOGRIE ID
Age: 26 years
Weight: 95 kg

6 The crew and the three Vogans are ready to return to the Endeavour.
There is a fault in the Transporter. They receive the following message:

TRANSPORTER DISKS INSTRUCTIONS

THE MIDDLE DISK IS FAULTY –
DO NOT USE.

EACH VOGAN MUST BE PLACED IN A
DIFFERENT ROW AND COLUMN.

THE TOTAL WEIGHT IN ANY ROW,
COLUMN OR DIAGONAL MUST NOT
EXCEED 235 KG.

TRANSPORTER ROOM

Use the diagram and information on **Workbook page 45, question 2** to
show the positions of the crew and Vogans to ensure a safe trip back.

Rescue on Vog

1 On route to Quizard, Starship Endeavour receives this
coded signal from Queen Mayo on planet Vog.

DATE: 19 29 09 19 39 39   *TIME:* 88 49 88 49

MESSAGE: 76 46 27 67 / 17 86 47 66 / 17 06 47 18 57 /

          18 06 17 46 47 / 16 68 / 78 57 87 66 57 47 08 / 37 06 68 57

(a) Decode the signal to find the date,
time and message.

(b) Put this reply signal in code:

ON OUR WAY. WILL
BE THERE IN 8 DAYS.
COSMOS

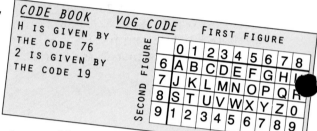

CODE BOOK   VOG CODE

H IS GIVEN BY
THE CODE 76
2 IS GIVEN BY
THE CODE 19

FIRST FIGURE

	0	1	2	3	4	5	6	7	8
6	A	B	C	D	E	F	G	H	I
7	J	K	L	M	N	O	P	Q	R
8	S	T	U	V	W	X	Y	Z	0
9	1	2	3	4	5	6	7	8	9

SECOND FIGURE

2 At a distance of 127 500 km from Vog, the Starship
Endeavour's port engine is damaged by a meteor.
The Starship then travels

     64 000 km in the first space hour

     32 000 km in the second space hour

     16 000 km in the third space hour and so on.

How long after the meteor struck did the
Starship arrive on Vog?

We beamed
a rescue party
down to Castle
Vog.

3 The route to the
Zorgon Palace where
the king is held is
shown by the dotted
line.

(a) The radio mast is
350 km due East of
Castle Vog. Find
the scale of the
map.

(b) Find the distance
and direction from
• the Mast to the
Fort
• the Fort to the
Well
• the Well to the
Palace.

(c) Describe the route
in your own words.

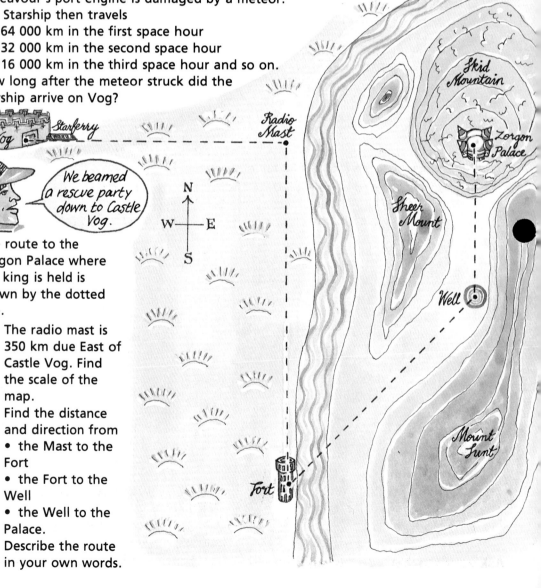

4

When we reached the Fort we discovered that acid from the swamp had damaged our Starferry beyond repair. We built a log raft to cross the river.

The rescue party allows two metres of rope altogether for knots and loops. What is the total length of rope they need?

5 The raft can carry a maximum of 160 kg.
Describe how the rescue party could cross the river.
Poisonous vapour rising from the river means that no one can make more than 2 return trips.

RESCUE PARTY	WEIGHT
CAPTAIN COSMOS	53 KG
LIEUTENANT DAZ	80 KG
MAJOR DAY	74 KG
DR SPLINT	87 KG

Remember I space hour = 100 space minutes.

Zorgon Palace

6 Skid Mountain is 820 metres high and is covered in slime.
Lt Daz climbs 100 metres in 80 minutes and then rests for 20 minutes. While resting he slips back 10 metres.
How long will it take him to reach the top?

Skid Mountain

7 The rescue party and the prisoners escape from the palace in a damaged Zorgon starship. The steering is faulty.
The starship travels 100 km north, 200 km east, 300 km south, 400 km west, 500 km north, 600 km east, and so on.

  (a) On $\frac{1}{2}$ cm squared paper draw the path of the starship.
  (b) The castle is 600 km due west of the Zorgon Palace. How far will the starship have travelled by the time it reaches Castle Vog?

8

Orbit

Vog

Before we left for Quizard we orbitted in the Endeavour 50 km above the planet Vog

The diameter of Vog is 1500 km.
Calculate

  (a) the diameter of the orbit
  (b) the distance travelled by Endeavour in one complete orbit.

Ask your teacher what to do next.

Impossible shapes

You need 1 cm isometric paper.

1 This is an impossible triangle You can
draw it but it cannot be made.

  (a) Copy the shape.

  (b) Your impossible triangle has a side
length of 6 cm. Draw an
impossible triangle with a side
length of 8 cm.

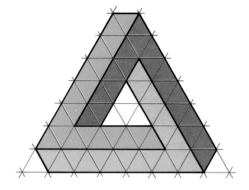

2 This is an impossible rectangle.

  (a) Copy the shape.

  (b) Your impossible rectangle
measures 8 cm by 4 cm.
Draw an impossible rectangle
which measures 7 cm by 5 cm.

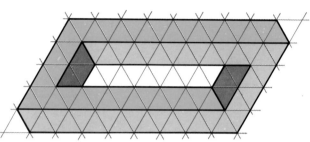

3 Copy and colour each of these
impossible shapes.

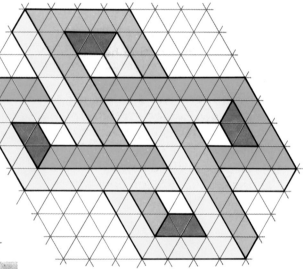

Truly Impossible
The artist M C Escher was fascinated
by impossible figures. Look carefully at
the water flow in his drawing
'The Waterfall'.
You can find more impossible shapes
and intriguing ideas on how to create
your own in the book *Adventures with
Impossible Figures* by Bruno Ernst.

Ask your teacher what to do next.